CLIMATE AND ECOSYSTEMS

Princeton Primers in Climate

CLIMATE AND ECOSYSTEMS

David Schimel

PRINCETON UNIVERSITY PRESS   *Princeton & Oxford*

Copyright © 2013 by Princeton University Press

Published by Princeton University Press,
41 William Street, Princeton, New Jersey 08540
In the United Kingdom: Princeton University Press,
6 Oxford Street, Woodstock, Oxfordshire OX20 1TW
press.princeton.edu

All Rights Reserved
ISBN 978-0-691-15195-3
ISBN (pbk.) 978-0-691-15196-0

British Library Cataloging-in-Publication Data is available
This book has been composed in Minion Pro and Avenir
Printed on acid-free paper. ∞
Printed in the United States of America
10 9 8 7 6 5 4 3 2 1

This book is dedicated to my student, colleague, and friend, the late Elizabeth Sulzman, a pioneer of Earth System Science and an inspiration to many.

Contents

Preface

THE GOAL OF THIS BOOK IS TO PROVIDE A BRIEF INTRO-
duction to the role of ecosystems in the climate system, and
in the changing Earth System. The climate we experience
on the surface of planet Earth results from interactions between *spheres* with different compositions and behaviors.
Earth derives the bulk of its energy from the sun, and so
solar–terrestrial interactions control the amount of energy
for life. The atmosphere continuously exchanges water
and energy with the oceans, and so *ocean–atmosphere*
interactions govern many climate phenomena, as well as
life in the oceans. Additionally, scientists have recently
discovered ways in which organisms affect the climate,
through myriad pathways. These *atmosphere–ecosystem*
interactions are the focus of this book.

This book covers the effect of climate on the behavior
and physiology of organisms, on the geography of organisms, and on the distribution of species of plants and animals over the planet. The text discusses how organisms
influence the exchange of matter and energy at the land
surface, and how terrestrial and marine organisms affect
the composition of the atmosphere and hence its energy
balance. By considering how interactions ecosystems
affect the atmosphere, we can analyze how ecosystems
cause changes that will affect the future climate.

This book is a broad overview of climate–ecosystem interactions. It covers terrestrial and marine ecosystems and addresses the organismal and physical/chemical aspects of ecosystems. Many excellent reviews of specific aspects of ecosystem–climate interactions are available (see references) for readers who need more detail on particular problems. Climate change is a major theme of the book. It provides a foundation for understanding how ecosystems interact with climate change ultimately caused by fossil fuel burning and describes mechanisms through which ecosystems change in response to climate change.

Climate and Ecosystems complements the other books in the Princeton Primers in Climate series that cover the physics of the Earth System. The responses and effects of the biosphere are intimately tied to the physics and chemistry of the climate system, and although this volume stands on its own, the reader will find greater depth on many phenomena in the other Primers in Climate volumes that will further illuminate issues covered in this book: the whole of the series of primers is more than the sum of the separate volumes.

Climate and Ecosystems is organized as follows:

Chapter 2 covers the basics of climate and energy balance from an organism's-eye view and provides critical physical concepts that underpin the rest of the book.

Chapter 3 addresses the effects of climate on ecosystems, including geographic patterns and the physiological and behavioral responses of organisms.

Chapter 4 focuses on ecosystem effects on climate through control of atmospheric chemistry, the carbon cycle, and the surface energy balance.

Chapter 5 discusses challenges and progress in modeling the future of ecosystems as the climate changes, and the interactions between ecosystem change and the climate.

Chapter 6 is a concluding discussion of climate change and some of the scientific frontier issues that form a basis for informing society about risk and opportunity.

This book is a product of my own curious path. I began my graduate career interested in the role of herbivores in ecosystems, particularly in their role in biogeochemical cycles. Through studies of nutrient loss to the atmosphere from urine and feces (really) I gained experience in trace gas flux measurements and modeling. When the much broader role of biogenic gases in climate and ozone depletion emerged (chapter 4), my trace gas skills became relevant to a new class of problems that coupled ecology and atmospheric chemistry. I gained a new world of colleagues in atmospheric chemistry and climate science and eventually moved to the National Center for Atmospheric Research to pursue studies of nitrous oxide and other nitrogen gases in the atmosphere. This work led to a great collaboration with Andi (Meinrat) Andreae and one of the first books on atmosphere–ecosystem interactions.

My involvement with climate studies began in a random way. In the mid-1980s my friend and colleague Bill

Parton and I did some consulting work for Budweiser Breweries. The company was building a brewery in Fort Collins, Colorado, and wanted to dispose of beer waste by spraying it on farmland. We were engaged to evaluate the impacts of that process on soil nitrogen. Our colleague John Altenhofen persuaded us to use our earnings to buy an integrating light sensor (a LiCOR ceptometer), an instrument that could be used to estimate light interception by plant canopies. He thought this was an emerging approach for estimating productivity (see chapter 3).

Soon after, Bill and I read a NASA call for proposals for a large climate and ecosystems field campaign that emphasized light interception and the emerging LUE paradigm (see chapter 3). We had never focused on this aspect of production ecology, having worked mainly on nitrogen limitation, but, hey, we had a ceptometer, so we could measure canopy light use. We were successful, and as a result of that project, the First International Land Surface Climatology Project Field Experiment (FIFE), I began working directly with climate scientists and other researchers like Piers Sellers who were pioneering coupling the land surface with climate models. FIFE led to many more projects and a lifelong involvement with climate science, remote sensing, and biophysics. I have often reflected on the consequences of that ceptometer purchase and my debt to the Fort Collins Budweiser Brewery.

In the mid-1990s I was asked to lead the first Intergovernmental Panel on Climate Change assessment of the carbon cycle, which recognized that ecosystems are at the heart of many of its unknowns, and I began a rapid learning process that eventually engaged me deeply in understanding the entire carbon cycle and not just its terrestrial components. The emerging techniques of inverse modeling and data assimilation for understanding land and ocean fluxes were exciting and related well to my long-term interest in modeling techniques. Using the atmosphere as an integrator, as is discussed in chapter 4, allows understanding the global carbon cycle but is limited in its resolution of specific processes and regions because atmospheric mixing blurs many of the details.

In the same period, rapid advances were occurring in the science of remote sensing of the biosphere using satellite-borne sensors. These new measurements complement atmospheric measurements by providing a global view, but one that is resolved in time and space. I became extremely interested, again, because of the work that began with the ceptometer, along with many colleagues, in using satellite-based techniques to understand where and why carbon exchange was changing. This period, while NASA's Earth Observing System (EOS) was being planned and launched, was a time of amazing discoveries, hard-fought scientific controversy, and great technical achievement, and laid the foundation for the extraordinary developments occurring today. It was a time of all-night meetings, early-morning conferences, huge decisions, and adrenaline-charged science, different from the pace and scale of most ecology.

In the first decade of this century, the pace of climate change, with its impacts on ecosystems, has been far faster than I or most colleagues expected. The investment during the 1990s in observing systems resulted in the detection of changes to growth patterns, phenology, species ranges, and the carbon cycle that hadn't been expected until perhaps midcentury. For a variety of reasons, most scientists underestimated how rapidly ecosystems might change. Models, it turned out, consistently underestimated the physical and biological sensitivity of the Earth System. I wrote this book with a growing sense of urgency about the need to understand—but even more important, to begin to manage—the climate and its impacts on the biosphere and humanity before irreversible and damaging change is inevitable.

The authorship of a book like this never does justice to all the people who contributed. This book is for my wife and best field companion, Susan Bonfield, who inspires me and has taught me to care about and focus on organisms and not just abstract systems. I'd like to thank my mentors, Jerry Melillo, Bob Woodmansee, Bill Parton, Vern Cole, Berrien Moore III, Inez Fung, and Francis Bretherton, and my many friends and partners in learning about the Earth System, but especially Rob Braswell and Scott Doney, to both of whom this book owes a great deal.

I have benefited from the best of colleagues, Arvin Mosier, a great trace gas and isotope scientist; biogeochemist Andi Andreae; biogeochemist Pam Matson; Dennis Ojima, companion on many a scientific venture; Piers Sellers, astronaut and biophysicist par excellence; Russ Monson, plant physiologist and global ecologist, tropical ecologist, and NEON partner; Michael Keller; Steve Running, cowboy global tree physiologist; Tony Janetos and Peter Backlund, colleagues in assessing and communicating climate change impacts; soil oceanographer Sue Trumbore; and Chris Field, global ecologist and skiing companion. This list doesn't begin to acknowledge all the scientific debts I owe to students, postdocs, collaborators, and colleagues, and I thank them all.

The view of science as a pyramid—of standing on the shoulders of giants—suggests a linear and hierarchical process. To me, research seems more like standing in a crowd, drinks in hand, passing ideas passed throughout the room. If you are lucky, when you wake up the next morning, one of those ideas takes hold and you can run with it. The next time you're with colleagues again, you toss out your morning idea and see what happens.

CLIMATE AND ECOSYSTEMS

1 INTRODUCTION

THE PHYSICAL ENVIRONMENT
THAT ORGANISMS INHABIT

THE INTERACTIONS BETWEEN CLIMATE AND ECOSYSTEMS occur on different timescales—a day, a year, or longer. On short timescales, we refer to *weather,* the actual atmospheric sequence of events (storms, wind, daily temperature). *Climate* is defined as the average of these events over years and longer time periods, and is described by storm frequency and intensity, mean wind, average temperature, and so on. Ecological systems experience and respond to atmospheric events (weather) as well as change more slowly with average conditions (climate).

A DAY

Consider a single summer day in a forest. As the sun comes up and temperatures warm, trees become active and the chemistry of photosynthesis begins. Some animals begin their daily activity, while others may seek refuge. Over the course of a single day, trees transition from removing carbon from the atmosphere and growing (daytime photosynthesis), to ceasing photosynthesis (nighttime) but releasing some of the day's photosynthetic gain of carbon back to the atmosphere during continued metabolism. During photosynthesis, water evaporates from the leaves, cooling not only the leaves but also the air above the forest. While trees are growing, changes in their size are usually imperceptible over a single day.

A YEAR

Let's expand our perspective to a year. In the spring, leaves begin to grow and expand, drawing on energy stored the previous year. The plants begin to take up carbon from the atmosphere, but the ecosystem as a whole still mainly respires stored carbon back to the atmosphere. As we'll see later, the atmosphere records this annual cycle of photosynthesis and respiration, so cumulatively, these plant processes affect the entire planet. As the weather warms and the season progresses, the carbon balance shifts, and photosynthesis begins to exceed respiration, leading to net growth of the biosphere. Leaves expand, plants increase in stature, and the daily cycle, described previously, continues within this grander cycle of the seasons. As winter and cold temperatures approach, growth ceases, leaves are shed, and respiration again exceeds photosynthesis.

A DECADE

Stepping back further, let's look at the ecosystem over a decade. Within the days and years of the decade, we see the preceding cycles, but we may also see a less orderly pattern. Some years have warmer or wetter conditions, while others are cooler or drier. The vegetation follows these climate signals, with more growth one year and less another. During a particularly dry year, a fire may remove most of the plant growth present, leaving bare soil. The carbon the forest stores or loses to the fire affects the amount of carbon in the atmosphere and eventually affects the climate. We begin to see the first clues as to how climate *change* may affect ecosystems, as variations in the physical resources plants need to grow (water, sunlight, heat) cause variations in the growth of individual plants.

A CENTURY

If we observe over many decades, we may note that although rainfall varies from one year to the next, the *average* amount of rainfall is changing. Although all the trees in the forest grow more in wetter years and less in drier years, some species are affected more than others. The more drought tolerant trees grow faster, and they may come to dominate the forest, inititating a change in its *species composition*. Thus, the effects of climate on individual forest organisms begin to be translated into altered relationships among species.

Imagine a drying trend. The increasingly taller drought-tolerant trees begin to shadow the water-loving species and reduce the light available to them for growth. Even in wet years, the drought-tolerant trees now have an advantage, and the entire community of organisms (including the animals that feed on the trees, and the predators that eat those herbivores) begins to change.

Although climate initiated the change, the interactions among the forest's organisms now take hold and control some of the change. Did a water-loving tree die of drought, or did it perish from inadequate light for growth owing to shade from a taller drought-tolerant tree? Was this tree death a climate effect or an effect of plant community processes? Of course, it was both, and in this simplified tale we can begin to see the complexity of climate–ecosystem interactions. As the forest canopy grows and covers more of the landscape, it makes the land surface darker, so it absorbs more sunlight, warming the surface more, and actually begins to change the local climate. Climate affects the metabolism and behavior of individual organisms, but these biological changes affect an organism's interactions with other organisms, and both the physics and the ecology of the system.

THE GLACIAL CYCLE

Climate and ecological change over decades is difficult to perceive, and scientists are just beginning to understand it, but the climate system and life are coupled on longer timescales as well (Barnola et al. 1991). The glacial–interglacial cycles, during which the earth cools and allows the growth of huge ice sheets, and then warms, releasing the water stored in the ice back into the oceans, are familiar. Some of the biological changes that occur over millennia as the earth warms and cools are similar to those described for a century. Some species fail, while others prosper, in glacial or warm interglacial climates. On these timescales, changes to marine and terrestrial carbon cycles have significant effects on climate, acting as controllers and not just responding passively.

However, over millennia, evolutionary change also occurs. Within species, cold- or heat-dominant genes may become more or less common, and entire species may arise or become extinct. Some of these changes may occur because of direct effects of climate. For example, a species may be unable to adapt to cold conditions, and all members of that species may die or fail to reproduce. Cold may reduce the numbers of a key prey species, leading to the extinction of a predator, or another predator species may be better able to travel over snow and thus may drive a competitor species to extinction through its higher effectiveness in a snowier climate.

THE GEOLOGICAL TIMESCALE

Climate and life also change together on the longest timescales. Paradoxically, these relationships may be the most familiar, as we know, for example, that dinosaurs flourished in a warmer past epoch of the planet. On the geological timescale, species or entire phyla flourish and decline in synchrony with vast, slow changes in the climate. On these timescales, life affects the geological and geochemical Earth System, changing rates of erosion (as land plants developed, they anchored the planet's soils); weathering of minerals (by fixing carbon and releasing it in soils as acidic compounds that affect mineral chemistry); and changing the composition of the atmosphere, releasing oxygen, methane, nitrous oxide, and other chemically or climatically important gases.

THE HUMAN TIMESCALE: THE ANTHROPOCENE

Geological time periods reflect events that are recorded in the rock record, through volcanic or erosional processes and other events that leave global traces evident to geologists. Recently, scientists have discussed terming the present the Anthropocene, because the effects of human use of natural resources, construction of cities and other infrastructure, climate change, and the impact of human-caused mass extinctions on the future fossil record should be evident to far-future researchers. In trying to understand present climate–ecosystem interactions the impacts of humanity are crucial. Human activity can change the way events occur over many different timescales. Harvesting a forest can instantaneously remove most of the wood slowly accumulated over days to centuries. However, that removal resets the forest's clock and will influence its dynamics for—at least—the lifespan of those trees. Human disturbance (forestry, conversion to agriculture) tends to cause rapid change to ecosystems but triggers slow responses as systems recover biomass and species composition over decades.

Of course, living systems are responding on all these timescales simultaneously. Early efforts to understand climate and ecosystems took shortcuts and tried to identify dominant influences of one timescale or another, but now we know that all these processes interact on different timescales. Year-to-year differences in crop yield may be due to just one extreme weather event. Centennial changes can arise when a fire or drought resets the age structure of a forest. A long-term trend may drive ecosystems to a state in which they respond differently to an extreme event. In the western United States, long-term trends in forest management have changed the sensitivity of forests to drought by allowing thick stands of trees to develop in the absence of fire; when drought comes, the dense forests (which fully use all the water available in wet years) are more stressed than they would be if there were fewer trees. In marine systems, slow changes in climate may influence long-lived fish populations, again changing the vulnerability of the system to rapid changes in phytoplankton following a climate event such as El Niño. Examination of the contingent and interacting effects of events and processes on different timescales is a major theme of this book and, as we'll see, provides much of the interest, challenge, and complexity of this science.

This book discusses the role of the earth's living organisms in the Earth System (ESSC 1988), which comprises the interacting atmosphere, oceans, lithosphere (soil and rocks), cryosphere (snow and ice), and biosphere, all influenced by and, most important, ceaselessly interacting with human activities (see figure 1). The biosphere affects the other Earth System components and is, in turn, influenced by them in many ways. A few decades ago, most scientists thought that life exists within the geophysical Earth System but influences it only in minor ways. The reality is more complex and more interesting.

Ecosystems and their interactions with climate vary greatly in different physical regions of the planet. Warm

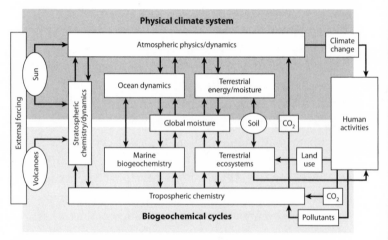

Figure 1. The Bretherton Diagram showing the key components of the biogeophysical Earth System and its human dimensions. Although the Earth System is considered a topic of interdisciplinary study, Bretherton pointed out that the Earth System, as an object, is also a subject for deep and coherent study. Many of its most important behaviors arise from the interactions among systems studied by the traditional disciplines (atmosphere–ocean, human behavior–atmospheric chemistry, etc.), requiring a new approach to research. (Source: ESSC 1988)

and wet climates have abundant growth and great diversity of organisms, and vegetation controls the flux of energy back to the atmosphere. Cold northern regions have simpler and less diverse systems but store vast amounts of carbon. The cold waters of the north have productive fisheries but lack the complexity and diversity of tropical reefs. While we observe that climate shapes life on the planet, great mysteries remain about how life responds to climate.

The living world, in turn, also shapes the physical and chemical Earth System. The composition of the atmosphere reflects the chemistry of life and is far from the chemical equilibrium that would obtain without the oxygen released by plant and microbial photosynthesis, the nitrogen converted by microorganisms into the forms that help warm our planet, and the water mined by trees from soils and released back into the atmosphere to cool the planet's surface.

The interaction of climate and life has been a scientific topic for centuries and an especially vibrant field of research for the past few decades. However, as the realization that our planet's climate is inexorably changing has dawned on humanity, understanding the effects of climate on living systems—and how life might affect the climate changes triggered by fossil fuel burning—has become more than an academic curiosity and is now needed to guide adaptation to these changes. Organisms, communities of organisms, and the great planetary biosphere itself respond to environmental change, and these changes affect the services the biosphere provides to humanity. While much is known about how the biosphere interacts with the rest of the Earth System, much remains unknown, This book focuses on how climate-triggered biological changes feed back to the physical and chemical parts of the Earth System across a wide range of timescales.

The scientific study of Earth is broken up into a number of disciplines, including atmospheric science, oceanography, ecology, geology, the natural resource disciplines of forestry and agronomy, hydrology, and—increasingly—the human studies including anthropology, history, economics, and geography. Often, subdisciplines concentrate on certain timescales as well, with paleoclimatology, paleoecology, and paleooceanography focused on the past. The human disciplines history and archaeology are distinct from studies of the present. You may be taking or teaching a course organized in one of these ways, but understanding climate and ecosystems and studying the Earth System draws on and unifies these approaches. This book is grounded in biology but draws on all these related approaches to studying our planet. The nexus of these different approaches is an emergent and coherent body of thought, sometimes called Earth System Science, and this book is written from that perspective.

2 THE CLIMATE SYSTEM

CLIMATE, CLIMATE VARIABILITY, AND RESOURCES TO SUPPORT THE LIVING WORLD

THIS CHAPTER FOCUSES ON THE RELATIONSHIP BE-tween physical variability in the environment and living systems. Organisms need *resources* from the physical world, including heat (to maintain body temperatures), light (to drive photosynthesis and for vision), water (to maintain hydration and solute balance), and chemical nutrients (which are linked to climate in complex ways we'll explore later). Organisms differ in their *needs* and *sensitivity* to these physical and chemical resources.

On short timescales, organisms respond to weather physiologically and behaviorally. Their growth rate may vary with temperature and incident sunlight, or their activity may vary between warmer and colder conditions. On longer climate timescales, animals may adjust their behavior and microhabitat selection or change their ranges. Scientists have studied these physiological and behavioral responses and the fascinating and unique ways organisms function in the planet's diverse climate zones.

Climate effects can be thought of as a cascade, from immediate and direct physical responses that cause physiological or behavioral reactions, to consequent but longer-term effects caused by changes in interactions among organisms in response to the direct effects. Organisms have to respond not only to Earth's diverse climate geography within each of the planet's climate zones but also to year-to-year variation and longer trends in climate within these zones. On these longer timescales (year to year and longer), organisms all respond to climate, but with differing sensitivities.

These differing sensitivities trigger changes in interactions among organisms (through competition, herbivory, predation, decomposition, and other processes). Short-term, direct responses produce a cascade of longer-term indirect effects as a result of interactions and feedbacks among the biota. This idea that immediate direct effects trigger complex and often surprising long-term consequences is a central concept of this book, which will be illustrated using the incredibly rich experience of researchers studying climate and life.

CLIMATE FOR BIOLOGISTS

For a biologist, it is critical to see climate through the senses and responses of organisms. Conventionally, we think of climate in terms of temperature, rainfall, sunshine, and wind as measured or forecast in very standardized ways. By contrast, organisms respond to the environment through the availability of resources: energy to fuel photosynthesis and metabolism, water to support hydration and cooling, and wind-driven motion that may transport mineral resources from great distances, which are all controlled by the processes we term climate. This section is a short primer on climate from the perspective of an ecologist. It discusses the way in which climate affects organisms and introduces the key variables used to describe those effects. Perhaps even more important, it discusses some of the temporal and spatial patterns of the *climate system* and how these affect living systems, and returns to the idea of cascades of effects on different timescales.

ENERGY BALANCE

The biological climate, or *bioclimate*, is largely defined by the flow of energy (Bonan 2008). Surface temperature is determined by the balance of incoming radiation, outgoing radiation, the evaporation of water, and the exchange of (sensible) heat through conduction and convection. The flux associated with the evaporation of water is one of the largest fluxes of energy in many ecosystems and is called the *latent heat flux*. This is also often the largest ecosystem flux of water. Latent fluxes occur when heat is used to evaporate water: energy is transferred to the water as it changes phase from liquid to vapor. Thus, the fluxes of water and energy are tied together in the surface energy balance, which is driven by radiation from the sun, quantified as net radiation. Net shortwave radiation is the solar shortwave radiation absorbed by the surface, after some is reflected based on the *albedo*, or

$$S_r = rS_i$$

where S_r is the net shortwave radiation; r is the albedo, or the fractional reflectance; and S_i is the incident radiation. The albedo of different Earth materials ranges from very bright (reflecting nearly all sunlight), such as snow (0.8–0.95), to dark, such as water (0.03–0.1). Vegetation typically has a low albedo, reflecting relatively little sunlight (0.05–0.25, depending on vegetation type and season). Longwave (thermal) radiation at the earth's surface, some of which is absorbed and some of which is emitted at a level proportional to the surface temperature is given by

$$L_e = \sigma \varepsilon (T)^4 + (1 - \varepsilon) L_i$$

where L_e and L_i are the emitted and incident longwave radiation, respectively; σ is the Stefan-Boltzmann constant (5.6×10^{-8} W m^{-2} K^{-4}); ε is the emissivity of the land surface, which is less than 1 (usually between 0.9 and 1); and T is Kelvin temperature. The *net radiation* is the amount of shortwave and longwave radiation absorbed and defines the amount of energy available to drive biological and physical processes at the land surface.

Some of this energy is transferred to the atmosphere as *sensible heat*, H, which is driven by the temperature gradient between the surface and air temperatures, modified by the resistance to motion of the surface:

$$H = \frac{\rho C_p (T_a - T_s)}{r_h}$$

where ρ is the density of air; C_p is the heat capacity of air (about 1005 J kg^{-1} °C^{-1}); T_a and T_s are the air and surface temperatures, respectively; and r_h is the resistance to flux, which depends on the height of vegetation, density of leaves, and other factors. Another mechanism, often the largest, is the transfer of *latent heat* or the transfer of heat in the evaporation of water. It is often computed using a similar resistance equation:

$$LE = \frac{\rho(q_a - q_s)}{r_w}$$

where LE is the latent heat flux; q_a and q_s are the actual and saturation specific humidities (kg kg^{-1}: kilograms water vapor per kilogram of air), respectively; and r_w is the resistance to water vapor exchange. From this equation we can see that evaporation of water depends on the driving force $(q_a - q_s)$, which is determined mainly by atmospheric humidity, and the resistance, determined by ecosystem structure and plant physiology. Some heat is also transferred to the soil via soil heat flux (soil heat flux, G; Bonan 2008). Combining all these terms, we define the surface energy balance by

$$(1 - r)S_i + \varepsilon L_i = \sigma\varepsilon(T) + H + LE + G$$

In detailed calculations, energy used in photosynthesis may also appear on the right-hand side of this equation. The left-hand side of this equation describes the influence of incoming radiation (the sum of absorbed solar and longwave radiation). The right-hand side describes the outgoing fluxes of radiation. This equation can be solved for the temperature that balances the incoming and outgoing fluxes via a complex, nonlinear equation. Models for this process have become more and more complex, from the Penman-Monteith equation (Monteith 1981) to modern land surface models (Bonan 2008) in which the resistances are modeled as complex functions of vegetation structure, control of leaf resistance (a component of r_w) by photosynthesis, and other biological dynamics.

CLIMATE VARIABLES

Net radiation, S_r, is a critical resource for living organisms, providing the energy needed for plant photosynthesis as well as controlling temperature. Incident shortwave radiation varies with season, latitude, and topography. The effective, or net, radiation also depends on albedo. The amount of light reaching the surface varies with weather (i.e., cloud cover), creating patterns that vary from day to day and year to year, superimposed on the more stable planetary patterns. Although light from the sun powers photosynthesis, when that light is absorbed, heat is produced. Plants must balance harvesting of light for photosynthesis against excessive heating of leaf tissue by regulating evaporative cooling. Although incoming solar radiation is much less frequently observed than temperature or precipitation, it is equally important for understanding how climate and its changes affect living systems.

Temperature is a key climate variable determining the rates of biological processes. While we typically think of *air* temperature as the index linking temperature and ecological processes, for an organism, the critical temperature is actually its *body* temperature, not the environmental temperature. An organism's temperature is determined by its energy balance (affected by the terms in the preceding equations), which is the balance between heat produced within the organism through its own metabolism and heat it gains or loses to the environment. The first term describing energy balance is simply the difference between the environmental temperature and the body temperature:

$$T_{env} - T_{org}$$

where T_{env} is the environmental temperature, and T_{org} is the organismal temperature. If T_{env} is greater than T_{org}, then the organism will gain heat from the environment and, conversely, will lose heat if T_{org} is greater than T_{env}. As with fluxes in the surface energy balance, the rate at which an organism gains or loses heat is affected by the resistance to flux and by the partitioning of the flux between sensible (H) and latent heat (LE). Insulation affects an organism's H, and LE is also controlled biologically (sweating, panting, transpiration). Temperature is normally measured and reported against a simple standard, developed for weather observations, based on a 2 m height, measured in an unshaded clearing. Climate models typically report surface temperature consistent with this standard as well.

However, the temperature an organism experiences is determined by its location in the ecosystem. Most organisms don't live at 2 m above the ground in a clearing. They live on a north- or south-facing slope (in the

Northern Hemisphere, south-facing slopes are warmer). They may prefer the sun or the shade. These local variations can be very large, and they determine the impact of climate. If air temperature is reported as 27 °C at 2 m from a standard measuring system, and an animal's body temperature is 31 °C, the organism should be losing heat to the environment $(27 - 31 = -4)$. However, the organism could be in a burrow, where the temperature might be 10 °C cooler.

To compute the effects of temperature or temperature change on this organism, the offset between the measured or modeled air temperature and the organism's temperature must be known. The difference is critical: if mean daytime temperatures were to increase at this location by 5 °C, then it would appear that the sign of the heat flux should change $(32 - 31 = 1)$. However, the actual difference in the animal's burrow would be $(22 - 31 = -9)$, suggesting a different thermal environment. Although relative temperature changes are important, absolute temperatures also matter.

This simple example reveals two important facets of temperature. First, for biology, the temperature that matters is the temperature where the organism lives, and second, the living world has many temperature thresholds: above a given temperature, one type of behavior occurs (e.g., gaining heat, and carrying out cooling behaviors such as panting, seeking shade, or swimming into a deeper layer in a lake), and below it, another type of behavior is present (e.g., losing heat, and seeking sunshine or a warmer layer). Consequently, a change in temperature must be given in terms of the starting and ending temperature ranges. A 5° change in temperature could increase or decrease thermal stress, depending on the current environment and a given organism's thermal tolerance. This means, for example, that warming in a mountain range, even if uniform with altitude, could increase stress at low altitudes and decrease stress at higher elevations!

Organisms have a wide range of mechanisms for managing high temperatures. Many of them involve water and the latent heat flux. The evaporation of water cools an organism, because energy is used to vaporize the water rather than to heat the source of the water. Thus, when a plant transpires water through its leaves, or an animal sweats, the water lost (the latent heat flux) cools the plant or animal. This phenomenon, along with the requirements of organisms to remain hydrated, links life to the water cycle.

Water is the next resource required by life. Water is most commonly indexed by precipitation, but precipitation is a very indirect measure of the availability of water to organisms. Organisms use water from soil and surface sources. In aquatic systems, precipitation controls critical features of habitat, depth, rate of flow, and temperature, but not directly. Water for rivers and lakes may come from afar, may have been stored as snow or ice, and may reflect the temperature of higher elevations. Precipitation defines the input of water to terrestrial and aquatic ecosystems, but the availability of water to organisms is also determined by how much of the water is lost as runoff, how much has already been used by organisms, and how much extra water enters through streamflow or even wind redistribution of snow. In a broad-brush sense, areas that have higher precipitation likely have higher biological availability of water, but the quantitative availability of water is a complex matter.

For vegetation, soil moisture is the most important index of water availability. Soil moisture is measured in only a few places and is often computed from the annual water balance:

$$\text{precipitation} = \text{runoff} + \text{runon} - \text{loss to groundwater} - \text{evapotranspiration}$$

Taking each term in turn, we define precipitation as the sum of rain and snow over some period of time (in some systems fog can also contribute). Runoff and runon occur in hilly and mountainous landscapes and depend on the slope and on the ease with which water infiltrates the soil. In forming groundwater, some water may infiltrate the soil below the rooting depth and become unavailable to the local biota.

Evapotranspiration is the sum of evaporation of water from soils or surface water bodies and the water transpired through the stomates, or pores, of leaves. Plants can control the aperture of their stomates to increase or decrease water loss from their leaves. CO_2 also enters through the leaves, so the wider the stomates open, the more water is lost, but the more carbon can be gained. Because CO_2 uptake requires solar radiation (sunlight), and radiation heats the leaves, and water loss cools the leaves, regulation of carbon gain, water loss, and leaf temperature are all closely coupled in plant metabolism to allow plants to maintain leaf temperatures at safe levels while maintaining growth.

In many terrestrial ecosystems, most water is lost by transpiration through leaves. Thus, water availability is regulated by the balance of precipitation inputs and plant-mediated transpiration. Therefore, plant water use is a major control over plant-available water. Obviously, precipitation provides at best an imperfect index of water availability, and observational or model-based indexes of precipitation provide only a rough indicator of water availability for life.

In fact, the coupling of precipitation to evapotranspiration creates a cascade of climate changes through ecosystem responses. Soil moisture may decline (or increase) over time because of changes in precipitation. In the short term, this will increase water stress on plants and may reduce carbon gain. If carbon gain declines, plant cover and leaf area will eventually decrease and thus will reduce evapotranspiration and change the relationship between precipitation and soil moisture (as less soil moisture may be used per unit time).

Since plant water use is an important pathway for returning precipitation to the atmosphere, the complex relationships among precipitation, soil moisture, and evaporation also affect surface water availability for animals, as does the topography. Steep landscapes, for example, offer many more possibilities for locally wet environments, where runoff accumulates, allowing for water-loving plant and animal communities in the midst of otherwise arid landscapes.

All precipitation is not equal, either! A given amount of rain or snow in a cold environment represents a larger resource for life in a cold environment compared with a hotter climate with high evaporation. Similarly, a given amount of precipitation is a larger resource on a flat or lowland site compared with a steep slope, where most of the water will run off.

Precipitation is only part of the story of water availability. Precipitation describes only the gross input of water to a region, which is then modified by evapotranspiration back to the atmosphere and redistribution within the landscape and soil column. We can understand how much water a given amount of precipitation represents only if we know the environmental temperature and humidity. The relationship between the amount of precipitation and the water resource for living organisms depends on the temperature (hotter environments are drier, other things being equal), the topographic setting, and other factors. Finally, water availability controls itself through feedback mechanisms. Because water availability and carbon gain are tightly linked through plant leaf processes, and because carbon gain ultimately influences the amount of foliage, on long timescales, changes in the water cycle influence vegetation and thus change the water cycle.

Wind is the poor cousin of the climate variables, and less is heard about it, but it exerts a significant influence on living systems. Organisms can lose heat to cold air near their surface. The removal of the warmed air by wind increases the temperature gradient between the organism and the atmosphere, increasing the rate of heat loss. This mechanism is the basis for wind chill: strong wind not only makes it feel colder but also accelerates heat loss. Similarly, wind removes air enriched in water by evaporation or transpiration. High winds increase water loss and increase the cooling effect of evaporation. Wind speeds vary greatly within an ecosystem, generally decreasing from plant canopy tops toward the soil. Again, computing the effect of wind on an organism's energy balance requires knowing not only an average wind speed for a region but also the variation in wind within the ecosystem. Extremely high winds also affect organisms as well, by toppling trees, causing flooding in coastal areas, and directly influencing the structure of living systems.

In the ocean, wind is a critical factor. The ocean tends to be stratified, with limited exchange of matter and energy between the layers (Vallis 2011). Stratification occurs because density is dependent on temperature. Since the ocean is heated from the top, layers become increasingly dense and colder at depth. This means that organisms living at a particular depth lack access to resources (nutrients or prey) at other depths. Wind provides energy for mixing in the upper ocean, and so the geography of wind speed is very important for marine life. Much of the ocean's circulation is wind driven. Offshore winds drive seawater away from the shore, and deep water must replace it. Thus, cold, nutrient-rich water from deep in the ocean replaces the surface water, whose nutrients have been depleted by active biological activity. Prevailing off-shore winds, coupled with other physical processes, can create rich zones for marine life by facilitating constant replacement of nutrients. When ocean conditions in the coastal zone change, major changes to the biota can follow, as is discussed in the section on El Niño (Vallis 2011).

CLIMATE PATTERNS

The temporal and spatial variation of climate shapes living systems. It is important to understand how the different climate variables affect and interact with living systems, but it is also critical to understand the patterns of variation in these variables (temperature, precipitation, radiation, wind, etc.). They are not independent of each other and tend to vary together in time and space. A full treatment of global climatology is outside the scope of this book, which focuses on a few aspects of the field critical for understanding climate and life (see Randall 2012 for a review).

Global climate patterns result from a number of factors. The equator receives more energy from the sun than do the poles and so is warmer. The energy contrasts between the poles and equator set up global patterns of atmospheric circulation, because air warmed in the tropics rises and once it cools, descends. As the heated air cools, its capacity to hold water decreases and thus is responsible for heavy precipitation in tropical regions. When the air finally descends, it is dry, and deserts tend to form just outside the tropics in the areas affected by the dry descending air.

Patterns or *climate regimes* are created by the arrangement of continents and oceans on the planet. The oceans store heat efficiently (water has a higher heat capacity and lower albedo than does continental rock and soil). In addition, the ocean can store heat in its mixed layer, the surface water that is sufficiently stirred to be of constant temperature. On land, heat must penetrate into soils and rock by much slower diffusive transfer processes. In summer, uptake of heat by the oceans tends to cool maritime regions, and in winter, the release of heat by the oceans moderates winter temperatures. As a result, temperatures in maritime regions are more stable, and the seasonal temperature cycle is less extreme. By contrast, the interiors of the continents are less buffered and so tend to be hotter in the summer, and colder in the winter, than coastal areas at similar latitudes. Broadly, rivers and lakes within the continents follow these patterns as well, modified by their depth, the temperature of their source waters, and other factors. The oceans also affect climate change: additional heat trapped by increasing greenhouse gases is efficiently absorbed by the oceans and so the air above the seas warms more slowly than the air above the continents.

Critical patterns are induced by mountain ranges (see figure 2). Obviously, higher altitudes tend to be colder and to have colder water in their rivers and lakes. River water of mountain origin may create cold-water habitats downstream as well. Mountains influence precipitation: as air rises over the mountains, it cools and can't hold

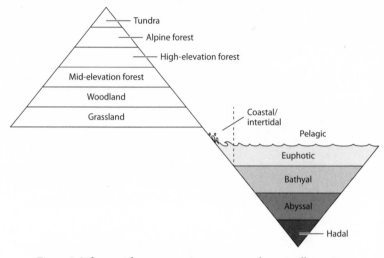

Figure 2. Life zones from mountaintop to ocean bottom, illustrating the similar and contrasting zonation of the terrestrial and oceanic domains. Temperature and precipitation cause most terrestrial vertical zonation, whereas the ocean is driven more by changes to light, chemistry, circulation, mixing, and temperature.

water vapor as efficiently, so mountains tend to receive more precipitation than surrounding lowland areas. The upwind sides of mountain ranges, which receive moist air first, tend to be wetter than the downwind sides, since much of the moisture was lost as the air first rose. Consequently, *rain shadows* are created. The effects on living systems of the patterns described are well known and are the basis for linking climate to *biomes*, that is, regions of generally similar vegetation and other organisms associated with particular climate regimes, such as deserts, rainforests, grasslands, and tundra.

Depth is also an important variable. Biologically significant patterns occur in the oceans, in three dimensions. Life in the open oceans, as on land, depends largely on photosynthesis, which occurs only in the upper, sunlit portion of the oceans. However, growing organisms in the surface ocean rapidly deplete nutrients, also needed for life, and these must be replenished from deeper layers. As mentioned previously, the density of seawater is a function of temperature, and also of salinity. In polar regions, when sea ice forms, the salt is expelled, resulting in unusually salty, dense water. This salty water sinks and is a driving force for global patterns of circulation (Vallis 2011).

Warm water from the tropics flows north to replace the sinking water in currents like the Gulf Stream, linking the oceans through global systems of currents (Vallis 2011). Much like on land, climate (light and temperature) together with patterns of low or high nutrients creates biomes or zones of life with distinctly different characteristics in the oceans. When climate varies over years to decades, these patterns can change, and the distribution of organisms in the oceans responds. Specific groups of organisms reflect marine conditions closely. Indeed, much of the geological record of change comes from preserved shells of organisms layered on the ocean bottom and so tightly linked to circulation and climate that these can be reconstructed from such organisms.

Whereas climate is described by the long-term statistics of the climate variables, its variation over time is often orchestrated into a series of modes, or characteristic patterns of space–time variability. The best known of these is the El Niño pattern, which occurs on a 3- to 7-year cycle. During normal years, the prevailing westerly winds force warm surface water to the western Pacific, allowing for upwelling of cold water along the western coast of South America. As a result, the western Pacific is substantially warmer than the eastern and supports intense atmospheric convective activity and heavy precipitation over the warm waters. During El Niño years, the winds weaken, and warm water flows eastward. These atmosphere–ocean interactions produce global climate effects. Most of the variation in tropical temperatures worldwide is a result of the El Niño cycle, and El Niño years are unusually warm.

The El Niño pattern produces unusual precipitation in many parts of the world (see figure 3). Heavier-than-average rainfall occurs over the tropical Pacific, but droughts in parts of South and Central America are common, as is drought in the monsoon region of Asia. Other regions, such as southern South America may have higher-than-average precipitation. We know that the El Niño cycle has strong reverberations in the biosphere, affecting the carbon cycle, wildfire frequency, reproduction in a number of tropical plant species, and even the frequency of some infectious diseases. There are many other important modal patterns in the climate system, such as the Pacific Decadal Oscillation, which is an El Niño–like pattern but over decades rather than years, and the North Atlantic Oscillation, which produces decadal climate variation in the eastern United States and Europe.

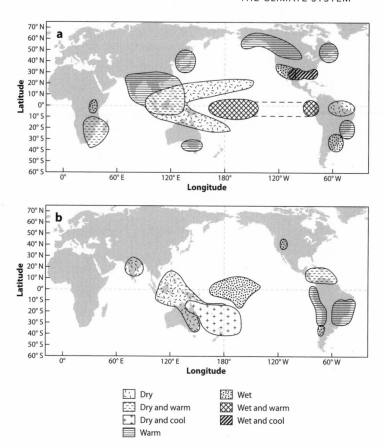

Figure 3. Impacts of the El Niño, a tropical Pacific climate process, are felt globally through changes to atmospheric circulation and the resulting redistribution and transport of water and energy. (a) Warm-episode (El Niño) correlations of the El Niño–Southern Oscillation (ENSO) climate pattern to temperature and water balance, December–February. (b)Warm-episode correlations of ENSO to temperature and water balance, June–August. Cool-episode (La Niña) relationships are approximately the opposite of these.
(Source: NOAA)

Climate variation tends to occur in organized patterns and does not affect the globe uniformly (although El Niño years tend to be warm on the average, some regions are typically cool in El Niño years). Climate variations tend to be organized patterns of all climate variables: El Niño affects temperature, precipitation, winds, and incoming solar radiation (via cloudiness).

Finally, modal variation also produces correlated patterns of biological response. Because of connectivity via immigration and emigration, the population dynamics of species may be affected differently when climate varies across the entire range, as opposed to when it varies only in some regions. Modal variations are known to have large impacts on fisheries, causing coordinated changes in fish size and age distribution over large areas of the oceans. Coordination of reproduction or disturbance by modal climate variations can result in anomalies in age distributions and can affect systems for years or, in the case of forests, decades. Climate mode impacts are not only a critical aspect of the planetary coupling between climate and life, but they also provide some of the most direct information about how longer-term trends in climate may affect the biosphere, a topic discussed in the next chapter.

How do human activities change the climate? The energy balance of Earth is maintained by a complex set of interactions between incoming solar radiation and the atmosphere and surface of the planet, many of which are described in chapter 4 in detail. Incoming radiation from the sun reflects the sun's blackbody temperature and has a maximum in the visible. Emitted longwave radiation reflects the temperature of the planet and is predominantly in the longwave or infrared (see figure 4).

Greenhouse gases warm the atmosphere by absorbing outgoing thermal radiation and reradiating it in all directions, including back to the surface and the lower

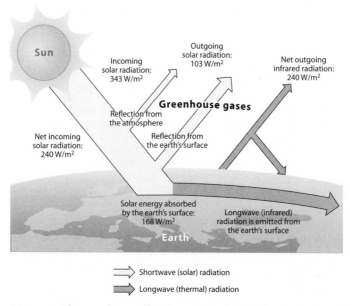

Shortwave (solar) radiation
Longwave (thermal) radiation

Figure 4. The greenhouse effect. Incoming solar radiation is dominated by the visible portion of the spectrum. When the incoming radiation is absorbed by Earth's surface, the planet reradiates in the infrared. Some of this infrared radiation is trapped by greenhouse gases (including water vapor), warming the atmosphere. Some incoming solar radiation is reflected as visible radiation by clouds, snow, and other bright surfaces and aerosols, and additional radiation is reflected from the land and ocean surface. The resulting temperature is a balance among the competing processes of absorption, reflection, and reradiation in the infrared.

atmosphere (Dickinson and Cicerone 1986). Greenhouse gases have always been part of this equation: trace gases such as water vapor, carbon dioxide (CO_2), methane (CH_4), and nitrous oxide (N_2O) warm the planet substantially. The warmer the air, the more water vapor it can hold and the higher the rate of evaporation. Earth is a water planet, so each increment of warming produced by greenhouse gases is amplified by the additional amount of water vapor a warmer atmosphere can hold. Water is also a highly effective greenhouse gas, so the additional water vapor further warms the atmosphere. However, the additional water vapor also affects cloudiness and can affect the planet's albedo (clouds are bright and reflect incident shortwave radiation) in different ways, depending on the type and height of the clouds (Randall 2012).

Although all these facts are agreed on, and based on theory dating to the nineteenth century, translating them into a quantitative prediction of the impact of changing greenhouse gases on temperature, precipitation, incoming solar radiation, wind, and other climate variables is very difficult. The best calculations from this theory estimate the global average, and so understanding climate change in terms of the spatial distribution of changes (mapping onto the global patterns described earlier) and in terms of space–time variability represents an extraordinary challenge.

Theory suggests a few details of spatial patterns of climate change. For example, the poles will warm faster than the rest of the planet, largely because as the polar weather warms, the amount and duration of snow and ice cover will decrease. Thus, a darker surface will be exposed, which means that less sunlight will be reflected, and more will be absorbed and reradiated as infrared radiation. Continental interiors warm faster than coastal regions. The oceans absorb more heat than the land, and so the air over the oceans, and land areas directly influenced by them, warm more slowly than the continental interiors. More climate extremes are expected, especially heavy precipitation. This *acceleration of the hydrological cycle* occurs because warming allows the atmosphere to hold more moisture and also increases the energy available for turbulence, both of which are conducive to severe storms (Randall 2012).

All these patterns, along with changes at local scales, have implications for impacts on ecosystems. Although warmer temperatures mean the atmosphere, as a whole, will contain more water, it will not be distributed evenly. Current patterns may be accentuated, so that dry regions will become drier, wet regions will become wetter, and more rainfall will occur in intense events (severe storms). Warmer temperatures and more severe storms increase the probability of direct climate effects on ecosystems in the form of drought, wind, flood, and other weather events whose frequencies may increase along with a warming climate.

When we consider the effects of a changing climate, we need to integrate all the scales of climate. We know that land plants and ocean plankton respond fairly quickly to extremes of climate, so slow trends may manifest themselves as changes in the probability of extreme climate patterns and modes. Some scientists have speculated that climate change will resemble a world with more frequent and intense El Niños. Most climate change scenarios suggest more frequent droughts, extreme temperatures, and heavy precipitation, coupling the long and short timescales of the climate system. The projection of climate change onto mode patterns like El Niño knits spatial scales together. The responses of many population and community processes to a climate extreme that affects all or most of a species range, synchronizing population processes across that range, are different from those produced by similar but spatially random extremes (Wang and Schimel 2003). Thinking across scales is challenging, and often crosses the disciplines scientists use to organize their thinking, but also reveals new organizing principles and simplifications, some of which we'll see used in the next chapter.

3 CLIMATE CONTROLS
OVER ECOSYSTEMS

....

CLIMATE AND THE DISTRIBUTION
OF ECOSYSTEMS

ONE OF THE OLDEST OBSERVATIONS OF ECOLOGY IS THAT climate and vegetation have corresponding patterns. Repeating patterns of plant structure and function occur in similar climate zones worldwide, and similar adaptations appear to have arisen in many lineages, suggesting evolutionary convergence to a limited set of biological solutions to environmental challenges. Ecologists have long known that climate plays a major role in the geography of ecosystems, and study it in the discipline known as *biogeography*, but they have struggled to develop a comprehensive theoretical framework to explain the patterns observed in nature. For decades, ecologists have studied the links between climate and the distribution of biological communities using geographic data and the paleorecord.

The biota are often described based on their common biological structure (forests, grasslands), and these structural types are often called *biomes*. Biomes are often described in terms of the dominant photosynthetic organisms (plants on land, or phytoplankton at sea). This

....

tradition includes early work on land describing the climatic controls over the global distribution with elevation or continentally of deciduous versus coniferous forests, grasslands, shrublands, deserts, and tundra systems. Biomes have been shown to correspond to patterns of precipitation, temperature, solar radiation, or the seasonality of these climatic controls (see figure 5). The Holdridge system is perhaps the best known. Using temperature, precipitation, and potential evapotranspiration, the Holdridge system classifies biomes worldwide into 38 types. The boundaries among these types can be described mathematically and can be incorporated into numerical models (Emanuel, Shugart, and Stevenson 1985). Elaborate work has been done relating boundaries of observed vegetation types (forests or grasslands) to subtleties of the climate (see chapter 5). Aquatic communities reflect climate patterns but are also influenced strongly by related factors such as the seasonality of flow in rivers or depth in lakes, the chemistry of the water, and the overall depth of lakes. Similar work in the marine realm has related differing groups of phytoplankton to variation in climate as well as to patterns of chemistry, including nutrients, silica, and micronutrients.

As ecologists explored the living world, correspondence between repeated patterns of climate and traits of biomes, communities, and organisms emerged. Once correlations between macroscale patterns of climate and ecosystem structure were recognized (e.g., Merriam and Steineger 1890), more detailed work within these biomes showed relationships between patterns of climate

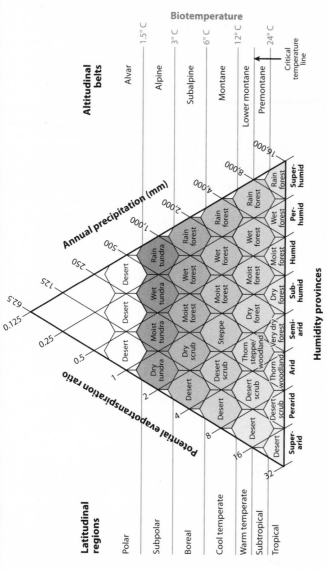

Figure 5. The Holdridge Life Zone System, which classifies regions according to biotemperature (mean annual temperature on a logarithmic scale), precipitation, and potential evapotranspiration. The system was conceived of as being three-dimensional but is normally illustrated in this triangular mode. (Source: Holdridge 1947)

and living communities within these ecosystems. As the scales of patterns became more detailed, the climate drivers evolved from annual-timescale patterns of temperature and precipitation to their modification seasonally or across landscapes. For example, Mediterranean ecosystems differ from other low-stature systems because they occur in regions of winter rainfall. Within northern forests, some communities tend to occur on drier uplands, where much of the precipitation runs off, and others in depressional areas where moisture accumulates. In these detailed landscape studies, the effects of disturbances such as wildfire and high winds became evident, but scientists often assumed that following disturbance, the ecosystem would follow a relatively deterministic trajectory back to the stable climate- and landform-determined state. This pattern of deterministic recovery after disturbance is referred to as *succession*.

Historically, ecologists have found clear climatic correlates to many species ranges. Some species are restricted to narrowly bounded habitats. Well-known examples are the salmonids, fish that generally flourish in cold water; warm-season or C4 photosynthetic pathway grasses that occupy hot and dry environments; and the many reptiles and amphibians that cannot survive cold.

Although many climate controls over species ranges seem intuitive (polar bears live in the Arctic, parrots in the tropics), some species may use an amazing range of environments. Cosmopolitan species like the coyotes range from subtropical to boreal habitats, and migratory birds that occupy very different habitats in breeding and wintering grounds have extraordinarily wide climate tolerance. For example, many North American thrushes and flycatchers breed in northern North America but winter in the tropics. Ecologists conceptualize these divergent types of environmental tolerances as forming part of a species *niche* (see figure 6).

There are a number of definitions of species niche. The basic idea is that the potential environments accessible to an organism can be described by a number of properties, such as minimum or maximum temperature, availability of a required food, pH, or a range of other environmental or trophic resources. These properties

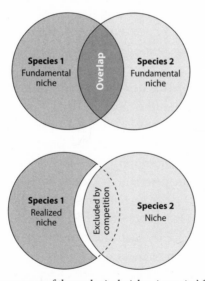

Figure 6. The concept of the ecological niche. A species' fundamental niche is determined by its environmental tolerance, but its realized niche may be narrower because of interspecific competition.

can be represented quantitatively as axes defining a multidimensional space determined by the environmental factors that affect the organism's evolutionary fitness (growth, reproduction). The niche maps population dynamics onto environmental factors and by doing so, links the organism's traits (physiology, life history) to its physical and biological environment. Since the organism's presence and abundance are linked to environmental factors, it should be possible to infer its niche from its spatial distribution, although observed ranges may contain unsuitable areas that are sustained by immigration.

Two types of niche are commonly considered. The fundamental niche is the entire environmental space inhabitable by an organism, that is, within which it can survive and reproduce (Holt 2009). However, in parts of that range, competition from other species may exclude the first species, so that its *realized* niche is smaller than its fundamental niche. For example, Bonan and Sirois (1992) evaluated the potential performance of black spruce, an important boreal forest species. They found, based on its physiology, that the spruce can grow well north of its northern range limit and that its growth optimum is coincident with its southern range limit. Presumably, its realized range limits (which lie well within its fundamental niche) are set by competition, rates of herbivory, or some other species interaction.

When climate changes, and species ranges are altered, the latter may be owing to direct effects of climate or to indirect consequences related to a food resource, a predator, or disease. Broad survey studies (Parmesan 2006) show that many species ranges are changing, and the range limits of many of these are moving farther poleward, suggesting a direct temperature effect. One well-known such system is that of the mountain pine beetle in the western United States (Powell and Logan 2005). This beetle has a temperature-limited life cycle, and a main cause of mortality when it is overwintering is a drop in winter temperatures below a threshold. Warming temperatures reduce the frequency of these killing cold snaps and allow beetle populations to grow from year to year. Thus, the range and abundance of the beetle is fairly directly controlled by temperature. However, the beetle is also a control over the range and abundance of several tree species, such as lodgepole and whitebark pine. Changes to the ranges of these trees may also appear correlated with climate, but the proximate cause of range shift is, in fact, beetle-induced mortality. Climate impacts on the trees are even more complex. The same warmer conditions that promote beetle outbreaks tend to lead to drier growing seasons. Trees are vulnerable to drought both directly and indirectly, since water-stressed trees have less resistance to the beetle. Changing temperatures are affecting the ranges of a large number of interconnected species in ways that have not yet reached equilibrium and may result in additional complications in the future.

Several assumptions have underlain much of the historical research on climate–ecosystem relationships. First, virtually all this research relies on the *equilibrium* assumption, an issue we will revisit repeatedly. By analyzing how the current distributions of ecosystems are related to present-day climate, we assume that those biological patterns are close to equilibrium with that climate. If we use relationships between observed distributions of biomes, communities, and organisms and climate variables, we are assuming those climate patterns determined those distributions. This assumption underlies a common modeling approach that assumes that if climate changes, ecosystem distributions will change in response to those climate changes. For example, in an early example, Emanuel, Shugart, and Stevenson (1985) related global biome patterns quantitatively to climate and used those equations to compute how biomes might move in a future world.

More recently, this approach is embodied in a group of models called species distribution models, climate envelope models, and Dynamic Global Vegetation Models (chapter 5). A wide range of models have been parameterized from the spatial covariance of biomes, communities, or organisms and climate, sometimes with additional biogeochemical constraints, especially for marine systems. The assumption of invariant relationships between climate and biology is often applied in paleoecology, in which past distributions of organisms (preserved in continental or marine sediments) are used to infer past climates. The equilibrium assumption will be discussed in more detail with respect to modeling species ranges.

The second assumption embedded in the classical climate–biology paradigm is that climate determines the distribution of organisms, and not vice versa. None of the classical work on vegetation and climate, the classics of biogeography, Clementsian or Gleasonian succession, or more recent individual-based vegetation modeling has allowed for feedbacks of vegetation to climate. Instead, most scientists have assumed that climate is a force extrinsic to ecosystems and that biology responds somewhat passively to those physical constraints. This is a useful simplification for some problems, but not correct, as will be discussed in the next chapter.

There is evidence for a more complex set of relationships and nonequilibrium conditions. For example, biogeographers have long known of the existence of *refugia*, or locations where populations of organisms persisted from previous climate eras. In some cases, these relict populations remained in sites with local climates, but in other cases, they persisted because of biological inertia, where stable communities effectively excluded immigration, for example, in bogs. Some of these refugia imply that biological communities have inertia, and at least some communities may persist into climates different from the climate in which they developed, demonstrating that not all communities are at equilibrium with present climate. However, first let's explore what we can learn from existing spatial patterns.

Beyond the distribution of ecosystem types, relationships between climate and ecosystems are determined from the spatial distribution of species—their ranges— and the correlation of species ranges with climate is fundamental to ecology. The impact of climate on ecosystems of the future will be determined largely by responses at the species level and by altered relationships among species. As a result of climate changes outside the ranges observed in the past, together with biological invasion and fragmentation, it is likely that new assemblies of biotic communities will not have direct analogs in the present or in the paleorecord. Climate affects the physiology of organisms directly and consequently affects food and habitat resources. Models for predicting these types of future communities are calibrated mainly from present-day (or paleo) species range–climate correlations.

The most familiar representations of species ranges are probably the maps found in any field guide, but the information inherent in species ranges is a central resource for the science of ecology. Current patterns of species and physiological trait distributions contain important information about functional relationships to climate and other environmental controls (Cody and Mooney 1978, Davis 1986, Bonan and Sirois 1992, Porter and Kearney 2009). The principal alternative to using range data to estimate species tolerances is direct measurement of physiological parameters, species by species (Bonan and Sirois 1992, Porter and Kearney 2009). Physiological studies are crucial, especially for species whose ranges are more constrained by competition with other species than by climate. However, to do this for all species is impractical in a world harboring about 250,000 plant species, and much larger numbers in other taxa, and the use of information inherent in spatial distributions is required. Distributional information is crucial for building models to understand future ecological responses to climate and other aspects of global change. The future planet will almost certainly contain novel species assemblages, so forecasting based on ecosystem types will break down, requiring modeling based even more strongly on species and trait characteristics.

There are many examples that show species range correlations with climate to be mediated by the temperature responses of forage or prey species. In the early part of the twentieth century, the biologist Joseph Grinnell surveyed the natural diversity of California in extraordinary depth. Between 1904 and the late 1930s Grinnell and his colleagues recorded the identity, location, and characteristics of tens of thousands of specimens and the habitats in which they were found. Joseph Grinnell was a pioneer of the niche concept and considered the niche to be the main control of a species' range. Grinnell considered the niche to be defined by the environmental conditions that limit a species to a geographical range where it can survive. He believed that temperature often ultimately defined range boundaries but was aware that numerous other factors, including interspecific competition, can restrict realized range boundaries. Grinnell's data are a valuable resource, because the detailed locations of sites and precise species-level information allow these sites to be revisited. Tingley and colleagues (2009) performed such a resampling, focusing on bird species. They wanted to know whether the distributions of these birds, which could presumably move easily in response to changing climate, were altered from those in the early twentieth century and whether those changes tracked climate in a predictable way. Birds are an ideal group, since they can move quickly and are less affected by barriers created by natural or human land cover change.

Tingley et al. used niche modeling techniques, also known as *species distribution models* (SDMs), to evaluate whether the new species ranges had tracked climate changes (temperature and precipitation) between the time of the original Grinnell survey and the present. They found evidence supporting a process, called *niche tracking*, in which species follow limiting environmental boundaries through geographic space to retain a favorable climatic space. As temperatures warmed, and precipitation patterns changed, the bird species Tingley et al. followed showed reduced populations or suffered local extinction in sites where the climate had become less favorable, and increased populations in areas where it was more favorable. While these observations suggest a strong and direct link between climate and species distributions, the details suggest more complicated mechanisms. In general, Tingley et al. found that high-elevation species, which live where temperatures are cold, adjusted their ranges to correspond to temperature. Low-elevation species, living in dry environments, appeared to respond to precipitation. Mid-elevation species showed sensitivity to both temperature and precipitation. Overall, more species showed sensitivity to precipitation, alone or in combination with temperature, than to temperature alone. These data suggest that climate was not the cause of the species range shifts, since precipitation tends to be a control over food and habitat rather than a direct control through a species' environmental tolerance. The range patterns suggest that productivity and, as a consequence, food may have been the dominant control. At low elevations, water limits plant growth and hence primary productivity, while at high elevation, temperature does. In the mid-elevation zone, both temperature and precipitation affect plant growth. The changes to ranges, though tightly correlated with climate, may have been due to underlying climate effects on productivity and energy availability for the birds, or even for the birds' prey (in the case of insect feeders).

This study, similar to that on the mountain pine beetle, suggests that species ranges provide important and usable information on climate effects. In the case of the beetle, temperature clearly defines one axis of its niche. Moisture appears to as well, but because moisture affects the ability of trees to defend themselves and not because precipitation directly affects the beetle. Similarly, precipitation and temperature can be used to describe the niches of many California birds, but the birds actually respond to climate through food availability and not because of, for example, their thermal tolerance.

If vegetation will respond to climate in the future much like it did in the past, then climate can be used to predict future avian species ranges. If, however, plant or insect productivity will respond differently than in the past, then the prediction based on past response will fail. For example, if crossing a temperature threshold allows an insect outbreak, greatly reducing productivity, then future ranges will no longer correspond to temperature.

This risk highlights the importance of underlying biological process. Although modeling the future based on historical ranges and data is a reasonable use of such data, mining the data for inference about process is crucial to identifying the likely robustness of predictions made using niche models.

Marine species, like terrestrial organisms, also have ranges, linked to physical factors in the ocean. However the physical factors are quite different. The independent or predictor variables in terrestrial environments are typically based on temperature and precipitation, and the seasonality and extremes of these variables. In one marine study the predictor variables included bathymetry (the topography of the ocean bottom), temperature, salinity, a group of variables related to oxygen availability, nutrient concentrations, and chlorophyll (Wiley et al. 2003). Chlorophyll is a correlate of primary productivity, and recalls the Tingley et al. (2009) conclusion that net primary productivity (NPP) may be the proximate control over distributions. The data also illustrate the differing measurements available. For example, nutrient availability is a control at the species level over many plant distributions, but nutrients are extremely patchily distributed in soils, whereas they are relatively well mixed in the ocean. Thus, while nutrients are a primary quantity mapped in the oceans, they must be inferred by modeling for terrestrial systems (Parton et al. 1987).

The aquatic and marine environments of species distributions are also shaped by the dynamic flows of water within bodies of water. This type of biogeographic control has no common analog in the terrestrial realm but can be critical in the ocean and larger lakes. Coastal marine range boundaries often occur where nearshore ocean currents collide. These currents often have very different physical and chemical characteristics and may contrast warm and cold water, or nutrient-rich with nutrient-poor waters. Marine biologists have assumed that the presence of marine range limits at water mass boundaries results from the strong water property gradients that occur along oceanographic discontinuities. Recently, however, researchers have begun to show that the flow regime itself may create biogeographic patterns (Gaylord and Gaines 2000). Common flow regimes (converging, diverging, recirculating) often observed along coastlines can constrain marine species' ranges, even if suitable habitat outside those ranges is available. This is true largely, although not entirely, because many marine species disperse primarily in their larval stages, when their independent mobility is limited. Current boundaries can function as barriers, either in one direction, or both, and create communities tied to the structure of the flow. This effect is stronger in stable, persistent flows, and weaker in more temporally variable current systems, and also depends on the life history of the organism. In particular, strongly swimming species could be less affected than passively dispersing species. Nonetheless, the fluid nature of the marine habitat creates additional structure beyond the types normally encountered in terrestrial systems.

How are marine species ranges linked to climate? Again, climate exerts a powerful influence over the marine environment, but in very different ways than in the terrestrial. Temperature is a direct control, and many commercial fishery species appear to be moving their ranges poleward or deeper toward colder waters (Doney et al. 2012). However, the relationships are complicated by the effects of water mass origin on temperature. The oceanic circulation transports heat, and, as described earlier, currents can intersect and have very different temperatures because of the origins of their water masses.

Global climate and ocean circulation are also coupled, producing relationships between climate and marine biogeographic regions and species distribution that may vary over seasonal to decadal timescales. The El Niño cycle greatly alters current in the Pacific and elsewhere, and shifts the distributions and abundance of many species dramatically. Other large-scale climate phenomena, such as the Pacific Decadal and Atlantic Decadal Oscillations (Hurrell and Deser 2009) have significant effects on ocean circulation, with strong consequent effects on marine life (Stenseth et al. 2002; Wang and Schimel 2003).

Because Earth System models still have relatively low resolution, their ability to resolve biologically important scales in the ocean is still evolving, and modeling of the complexity of physical-chemical-biological interactions in the future oceans remains preliminary. However, the dynamic nature of the ocean circulation, and the tight coupling between circulation, nutrient chemistry, and the growing impact of acidification have led to more emphasis on dynamic, process-oriented modeling rather than on the more static approach embodied in species range models. Projections from marine ecosystem models will be discussed in the final chapter.

In all cases, inferring ecological information from species ranges relies on a crucial assumption: the range must be near equilibrium. If a range is at equilibrium, then its limits identify the point beyond which some factor, climate or otherwise, will not allow growth and reproduction. The distribution of abundance within the range, and its correlation with environmental factors, provides some insight into how the population is regulated. In the Tingley et al. study discussed earlier, the species evaluated were assumed to be in equilibrium during Grinnell's surveys and at present. For birds, this is a reasonable assumption, since they are highly mobile. A wide range of adjustment times are possible, and species range data, by themselves, tell us little about these timescales. Given the long adjustment times possible, can we assume that existing species range data reflect equilibrium conditions?

The short answer is, with appropriate caution, we can. The present-day distributions of species and ecosystem emerged late in the Holocene period—the 12,000-year period ending with millennia of relatively stable climate conditions—during the latter part of which global mean temperatures varied by only a degree or so. While this is not long enough for full biotic adjustment to temperature, this period of extended stability allowed for the development of relative stable species ranges. Stable species ranges do not imply stable local populations, because ranges are a statistical estimate and may include areas where local populations have recently been established or just gone extinct for reasons unrelated to the species niche. For major types of woody plants, the response time after deglaciation was millennial in nature, limited by dispersal and community reassembly as well as by evolutionary responses. Later in the Holocene, evidence from atmospheric CO_2 and its isotopes—which are sensitive indicators of large-scale ecological change—support the idea that the global land biosphere and its large-scale carbon balance were relatively quiescent (Stocker, Strassmann, and Joos 2011; Elsig et al. 2009), and resulted in the trove of ecological data we find today in species ranges. It is worth noting that this trove is not secure, as high rates of land use and climate change will cause more and more species ranges to deviate from equilibrium. Thus, historical data on ranges will contain more information about species niches than data collected in the future (Schimel, Asner, and Moorcroft 2013).

NONEQUILIBRIUM PROCESSES

Forests are rarely, if ever, constant in time and are only statistically ever in an equilibrium condition. Cohen and Pastor (1997) approached nonequilbrium dynamics in northern forests using a modeling approach and derived a complementary result. They did not focus on migration rates but instead simulated the interaction of stochastic disturbance with climate change. They used a model that coupled plant growth and decomposition of dead plant material in soils: plant growth required soil nutrients, and decomposition processes in soils regulated the availability of nutrients. In their model, tree species were defined not only by their climatic envelopes but also by their nutrient requirements. Species that differed in nutrient requirements also produced foliage that decomposed at different rates, linking growth and decomposition processes, following the approach pioneered by Aber and Melillo (1982). When Pastor and Cohen exercised this model, they found that depending on the initial species composition and nutrient status of a given site, and the disturbance regime, climate did not determine a unique community composition but instead resulted in a chaotic array of possible trajectories of biological communities, all containing species consistent with the climate but varying because of the additional dependence on nutrient status, itself partly determined by species composition.

Pastor and Cohen, in linking community dynamics to species-specific nutrient requirements as well as climate, extended a paradigm originally developed in the oceans by Alfred Redfield. Redfield noted in 1934, and then in more detail in the 1950s, that the ratios of constituents critical to biology (carbon, nitrogen, phosphorus, silica) "in seawater occur in characteristic ratios (*Redfield ratios*) that reflect the demand for these nutrients by phytoplankton (Sterner and Elser 2002). Redfield hypothesized that living organisms had, over time, created a chemical environment for life in the oceans that was the result of biological uptake, incorporation into tissue, and eventual decay and sedimentary burial of organisms. This hypothesis has proved durable and powerful and still lies at the heart of marine biogeochemistry. Parton and his collaborators (Parton et al. 1987) incorporated this idea explicitly into a terrestrial biogeochemical model, and the idea that living organisms shape their geochemical environment has become widely accepted in marine and terrestrial ecology (Sterner and Elser 2002).

Crucially, if communities shape and are shaped by biogeochemical processes, then the climate–ecosystem paradigm takes on a new complexity. At least in terrestrial environments, organisms are significantly constrained by the availability of nutrients and by the redox environment defined by the geology and soils (Cole and Heil 1981). This implies that the climatic regulation of ecosystems plays out in a geochemically constrained theater. I argued (Schimel, Braswell, and Parton 1997) that even if the availability of some nutrients (such as nitrogen) also equilibrated with climate, the timescales for equilibration of nitrogen with climate were sufficiently long that systems might be in nonequilbrium states much of the time. Vitousek (2004) showed that in rock-derived nutrient-limited ecosystems (e.g., phosphorus) systems could be continuously evolving, albeit slowly, over geological time. In a single-climate environment, many biogeochemically different systems were possible, varying with soil age and depending on the degree of nutrient limitation and on the primary limiting element. Although all the sites Vitousek studied were recognizably tropical forests (and generally consistent with the climate), they differed very substantially in species composition, productivity, and diversity along gradients of geochemical age.

Marine systems provide an interesting contrast to terrestrial. In marine environments, species distributions are strongly influenced by the climatic and chemical setting. The equilibrium perspective is both stronger and weaker in the marine literature than in terrestrial ecology. It is stronger because the argument for equilibrium between physical/chemical conditions in species traits is stronger than in terrestrial environments. This is because most marine organisms, and especially photosynthetic organisms, are small-bodied with very short generation times. Species composition can change very quickly with environmental conditions, especially for phytoplankton and zooplankton. Also, terrestrial plants are rooted in soils that contain the accumulation of hundreds to thousands of years of dead organic matter, which creates a slowly changing biological and nutritional environment for vegetation and adds memory to the system. By contrast, the marine equivalents of soil are the deep nutrient reservoirs in the ocean thermocline and are only indirectly coupled to the photosynthetic ecosystems of the upper ocean. Whereas in terrestrial environments, long-lived plants and soil processes create barriers to rapid equilibrium between climate and ecosystems, in marine systems, short-lived organisms and spatial separation of photosynthetic and detrital-based ecosystems reduce such barriers.

However, marine ecologists have had to rely less on approximating environmental control by using (assumed) equilibrium patterns, because marine systems change rapidly enough that these can readily be observed. Marine systems change fast enough that community composition change can be observed by sampling over subannual time periods (for example, blooms) or during the course of short-lived experiments (for example, when a limiting nutrient is added to the surface ocean). As a result of these dynamic studies, biological oceanographers know that marine plant diversity productivity and biomass are controlled not only by physical and chemical conditions but also by herbivory and predation. Whereas terrestrial ecosystem models are dominated by representation of carbon, water, and nutrient and energy fluxes between plants and decomposers, marine ecosystem models include microbial decomposition and additionally focus on organic matter transformations mediated by herbivory and predation. Marine models produce much more complex, and much less apparently deterministic, relationships between physical/ chemical forcing and biological response. It remains to be seen whether such trophically driven processes are crucial to terrestrial biogeography. This debate continues in the literature.

CLIMATE AND ECOSYSTEM FUNCTION

Much of what we know about climate and ecosystems comes from observing and trying to explain the distribution and abundance of organisms, sometimes grouped into communities or biomes, and this knowledge is complemented by studies of physiological and behavioral responses to climate. While the subdisciplines of *ecophysiology* and *behavioral ecology* are somewhat younger traditions than biogeography, they provide mechanistic explanations to complement hypotheses derived from the patterns of correlation in space (or time, using the paleorecord). Physiologists and behaviorists are able to study such biological process in plants and animals as photosynthesis, respiration, thermoregulation, and predation, using either natural or experimental variation in climate to link these processes to the environment.

PLANT ADAPTATION TO CLIMATE

Both marine and terrestrial plants live tightly coupled to the environment. On land, plants exist in the atmospheric boundary layer and surface soils. Land plants directly transform environmental resources into biomass and energy for metabolism, coupling them tightly to climate. Light from the sun is the first resource to consider. Plants harvest incoming solar radiation in a complex cascade of physical chemical reactions in which solar photons drive reactions that convert inorganic carbon to organic matter. Although land and marine plants photosynthesize in a diversity of ways, the underlying biophysics are common to most contemporary plants. How, then, is light linked to climate, since it arrives at a nearly constant level determined by the sun, cloud cover, and the seasons?

The total amount of sunlight reaching vegetation is determined by the state of the atmosphere, and in the oceans, by the depth and turbidity of the ocean mixed layer. Clouds reflect sunlight back to space and diminish the amount of solar radiation reaching the vegetation.

The atmosphere contains a large number of aerosols, tiny particles dispersed in the air. Depending on their size and vertical distribution in the atmosphere, aerosols can interact with sunlight in a large number of ways. Most simply, they may scatter light, reflecting it in multiple directions, and so reduce the amount reaching the surface (because some of the scattered light will be backscattered toward space). These physical processes lead to variations in sunlight available for photosynthesis. First, different regions have very different cloud regimes. Many regions have diurnal cycles of aerosol and cloud cover, so the probability of sunlight may depend on the time of day. Clouds and aerosols can also vary seasonally, depending on the source of air masses reaching a particular location. Indeed, we often think of the seasons in terms of cloud cover, reflecting the probability of clouds in different seasons: "bright spring days" or "dull winter weather." Clouds and aerosols may also vary from year to year or have trends over years, producing interannual variations in sunlight, with consequent effects on biology.

Clouds and aerosols also affect the nature of the sunlight reaching the surface. In a clear atmosphere, sunlight reaches the surface as a direct beam of sunlight, producing bright, intense sunlight and dark shadow. The light clearly emanates from the disk of the sun. In a cloudy or aerosol-rich atmosphere, much of the light is scattered, and through multiple scattering may reach the surface from all directions. In the extreme, on a cloudy day, the light seems to come from everywhere and nowhere, and the sun's disk may be hidden. This *diffuse radiation* interacts with vegetation differently than does direct beam radiation.

On a clear day, when the sun's light arrives as a direct beam, those leaves that are in the sun receive high levels of radiation, but those out of the direct beam or in shadow receive very little radiation. Whereas the foliage in full sun photosynthesizes at high rates, those leaves out of the direct beam may not be photosynthesizing at all. In a location with diffuse radiation, with radiation arriving from all sky directions, most leaves will receive some radiation, although none will receive full sunlight. Which condition produces more total photosynthesis, clear days with maximum total radiation reaching the surface, or dull days with diffuse radiation penetrating and saturating the entire plant canopy? In other words, is the maximum photosynthesis achieved when a few leaves photosynthesize at high rates or when the entire canopy photosynthesizes at lower rates? The answer is, it depends. Maximum photosynthesis may occur under either condition depending on the total sunlight, the fraction of diffuse radiation, the structure of the plant canopy, and the species involved.

Globally, there is some evidence that years with high diffuse radiation (as occurs, for example, after the injection of massive aerosols into the stratosphere from volcanic eruptions) may have higher global photosynthesis, but this theory remains controversial. At any given location, though, there is no question that variation in direct versus diffuse radiation affects photosynthesis, as does variation in the total amount of radiation. Clouds and aerosols directly couple the climate system to vegetation and ecosystems, and produce time and space scales of variability that can be of great consequence.

Light is a crucial resource for marine photosynthesis as well. Although regionally and seasonally marine systems are often nutrient limited (that is, growth rates of primary producers depend on the supply of nitrogen, phosphorus, and micronutrients such as iron to the phytoplankton), light shapes marine systems in important ways. First, light is strongly attenuated vertically in the water column, producing an upper, sunlit or *euphotic* zone, where light is sufficient for photosynthesis to occur. The depth of this zone depends on many factors, including mixing depth and the abundance of marine photosynthetic organisms. Below the euphotic is the *aphotic* zone, where light is insufficient for photosynthesis or where rates are low enough that respiration exceeds photosynthesis. The aphotic zone is not without intense biological activity: especially when the euphotic zone is highly active, the deeper depths may have rapid and abundant detrital metabolism, producing carbon dioxide from organic matter and releasing nutrients.

In addition to vertical interactions with light, the diurnal and seasonal cycles of incident solar radiation and of cloud or aerosol cover can significantly affect sunlight available for photosynthesis in marine systems, producing seasonally varying patterns of nutrient versus light limitation, or colimitation. Unlike terrestrial systems, with their relatively stable vegetation patterns, the phytoplankton community can change very quickly, and species composition responds to the balance of limiting factors, leading to dynamic changes to primary producers and the food webs based on them. While most studies of land plants focus on their physiological response to changing light, in marine systems, community responses to changing limiting factors are as important.

After light, water is the next critical resource. All organisms are mostly composed of water and so need to remain hydrated to survive, or at least to remain active. Without a constant source of water, both plants and animals experience stress and mortality, although the requirements for water vary dramatically among species and environments. Many remarkable adaptations allow some species to survive and remain active in very arid conditions. For land plants, water provides a critical resource that links vegetation tightly to the climate system.

When photosynthesizing, land plants essentially trade water for carbon dioxide. How does this process work? Plants absorb carbon dioxide through tiny pores called *stomates* (see figure 7). The stomates can expand and contract, so their diameter regulates the rate of gas flow through them. When the stomates are dilated, gases flow freely through the pore. When they contract, the flow is restricted. When plants have abundant resources for photosynthesis, for example, when fully sunlit, the stomates tend to open to allow rapid flux of carbon dioxide toward the carbon-fixing pigments. When resources are scarce on cloudy or dry days, the stomates shut, restricting the flux of gases. It's easy to understand why

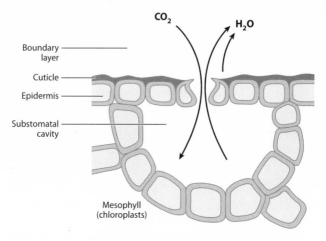

Figure 7. Cross section of a stomate—a microscopic yet globally important feature of plants. Stomatal control of water and carbon dioxide exchange has major impacts on the carbon cycle, the hydrological cycle, and Earth's climate.
(Source: Bonan 2008)

the stomates are open under sunny conditions. Why do they shut under cloudy or dry conditions? When the stomates are open to let CO_2 into the leaf, they also allow water out. The more CO_2 the plant takes up, the more water it loses to the atmosphere by transpiration. A plant growing in moist soil under full sunlight will open its stomates and photosynthesize vigorously; the same plant in dry soil, under the same full sunlight, will close its stomates to conserve water.

This apparently simple phenomenon of water and CO_2 exchange through small pores, couples the atmosphere, the biosphere, and the hydrological cycle globally. When water is transpired from the plant canopy (the array of leaves over a region), it cools the surface. The evaporation of water transfers energy as latent heat from the leaves to the atmosphere, cooling the surface. While the plant is trading water for carbon to fuel metabolism it is also using the water to maintain appropriate temperatures. Because of the central role of stomates in regulating carbon and water exchange, photosynthesis plays a central regulatory role in coupling climate and ecosystems. We discussed in chapter 2 how latent heat affects the surface energy balance. The latent heat flux is influenced by the resistance to water flux. One term in the resistance, r_w, is the stomatal resistance. The complex interactions that govern photosynthesis and transpiration thus also affect, and are affected by, the surface energy balance feeding back to the climate system. Other things being equal, high rates of photosynthesis imply high rates of transpiration, with concomitant impacts on the energy balance.

In the preceding discussion, light and water were considered as controls over photosynthesis. Many other factors also set levels of photosynthesis, and so they can influence water, carbon, and energy balance. These factors include soil-derived mineral nutrients necessary for making the complex biochemicals involved in plant metabolism, and atmospheric humidity, which influences evaporation through the stomates. The history of a given site can also influence photosynthesis. Following a fire, a system may have barely any foliage, so no matter what the conditions, photosynthesis and transpiration are inhibited. Management by humans also affects the carbon/water balance by controlling the species present, their nutrition via fertilization, and even the amount of water available through irrigation.

The coupling between resources such as water, light, and nutrients, and climate through water and energy exchange has led to important theories or models of how climate and biology interact, both of which employ analogies as part of their argument. A critical theory linking resource requirements and plant growth to evolution and individual fitness is referred to as the *economic analogy*.

THE ECONOMIC ANALOGY

As we have seen, plants must *acquire* environmental resources to grow: these include water, light, CO_2, and nutrients. Plants must *invest* resources to acquire resources, growing roots to acquire water and nutrients, leaves to harvest light, and stems to overtop competitors. Ecologists noted that these functions were analogous to those of a business firm and so they built a simple model based on principles of quantitative economics. The economic analogy is often useful in analyzing plant growth and modeling how plants should invest resources, especially when detailed empirical measurements are lacking. The approach is laid out in a paper by Bloom, Chapin, and Mooney (1985). The analogy assumes that the goal of plant growth strategies is to maximize primary production, or growth. The economic analogy is based on the following four theorems (expressed in words in

Bloom, Chapin, and Mooney, but all have mathematical counterparts):

Theorem 1: Plants should acquire resources at the lowest investment cost and use them when they are most valuable (for example acquiring nutrients in the spring, when they are abundant after the winter, and then using them midsummer, when light is most abundant).

Theorem 2: Plant productivity is optimal when the marginal cost (resources invested) of growth is equal to the carbon gain associated with that investment. To maximize NPP, a plant should continue to produce foliage until the increment of gross photosynthesis with new leaves equals the respiratory cost of those leaves.

Theorem 3: A plant should equalize the ratio of marginal growth over cost for all resources. This implies adjusting allocation between organs and processes so that for a given investment in acquiring each resource, the plant can achieve the same growth response. Optimally, growth is limited by all resources.

Theorem 4: Internal resources (for example, nutrients and energy) should be optimally allocated among competing plant processes (leaves, roots, stems, defense against pest and pathogens) until each resource limits each process to the same degree. Adding any additional resource should stimulate all processes contributing to growth to the same degree.

The economic analogy makes the following five predictions based on these theorems:

Prediction 1: Storage allows plants to acquire resources at minimal cost and use them at times when they have the maximum effect on growth.

Prediction 2: Plants will continue to allocate to organs (leaves, roots) or processes (photosynthesis, defense) until the gain in growth from investment equals the resource cost of the investment.

Prediction 3: Plants will adjust their allocation to organs and processes so that their growth is limited equally by all resources.

Prediction 4: Each plant process or organ is limited equally by internal allocation of resources.

Prediction 5 deals with timescales over which these adjustments should occur (theory predicts instantaneous adjustment, but this is not possible).

Prediction 5: Plants adjust in the short term (acclimation) and in the long term by genetic adaptation to optimize their resource use.

The economic analogy describes how plants should behave in the abstract. Numerous elegant studies have demonstrated the actual optimization of resource investment by plants, lending a degree of empirical support to the theory (Field 1983; Schimel et al. 1991), and it has been used with great effect in a variety of models (Sellers et al. 1992). Prediction 5 implies that when a plant reaches its limits of acclimation, competitive and evolutionary processes may come to dominate, leading to optimization by a succession of plant species or genotypes. The theory is extremely useful in simplified predictions of what plants should do in a given environment, especially since the mechanistic theory for explaining allocation of resources at a biochemical level is in its infancy, and few data exist for most species. The theory is also useful when it fails, as that demonstrates limitations to acclimational or adaptational flexibility and provides insights into the conditions under which abrupt ecosystem change may occur (as species exceed their maximal tolerances).

CLIMATE AND NUTRIENTS

In both terrestrial and marine ecosystems, carbon uptake and storage are greatly influenced by both physical factors and biogeochemical processes. Broadly speaking, physical factors determine the amount of energy available to the system through direct and indirect effects on photosynthesis. Biogeochemical cycles control the availability of nutrients for creating specific biochemicals important to photosynthesis, for other aspects of metabolism, and for building of structural tissue (wood, skeletons, and other types of structure). Because organisms need carbon and other elements in specific ratios, called *stoichiometric* ratios, and because organisms differ in their stoichiometry, understanding nutrient limitation is critical to understanding ecological processes. As we'll see, nutrient cycles interact with climate in fascinating ways. Marine and terrestrial organisms respond to limiting factors in similar ways, so that warmer temperatures and increasing light and nutrients generally boost photosynthesis and productivity. However, there are important differences between biological processes and interactions in the water column, and life in the soil–atmosphere continuum.

In the oceans, nutrients tend to be lowest in the euphotic zone. The abundance of light leads to rapid biological uptake of nutrients, and their incorporation into tissue. Although some of the photosynthetically fixed carbon in the euphotic zone is recycled through mortality and decomposition or herbivory and predation, much of it sinks below the euphotic zone. While photosynthesis does not occur in the aphotic zone, heterotrophic activity may be high, leading to remineralization of the organic matter and release of nutrients. Thus, nutrients tend to be vertically stratified in the ocean, with higher concentrations of CO_2 and nutrients at depth than in the surface water. Autotrophic (photosynthetic) and heterotrophic activity are thus separated by depth (both occur in the surface ocean, but heterotrophic activity takes place at depth) and linked by physical mixing. As we'll see, this separation produces important differences in the interaction of climate, nutrients, and ecosystems in marine versus terrestrial systems.

Because of this vertical stratification, photosynthesis and metabolism are high in the oceans when nutrients are supplied to the euphotic zone. There are many mechanisms through which this occurs, but most are linked to physical mixing and advection of subsurface water into the surface. Because transport drives nutrient availability in the oceans, the physical circulation of the ocean and its biology are tightly coupled. Ocean circulation is tightly linked to the climate system, being driven by winds, temperature, and salinity (density), and so marine biology is coupled with the climate system in ways that have no clear analogy in terrestrial systems. This section will conclude with some contrasts produced by this difference.

Transport of deep, nutrient-rich water into the euphotic zone (called *upwelling*) occurs for a number of reasons. One of the most familiar upwellings occurs off the western shores of South America, and at other eastern boundary currents. There, prevailing winds, in combination with the Coriolis force (motion due to the rotation of the earth; Vallis 2011) force the surface water westward, and so deep, cold, nutrient-rich water upwells to balance the westward motion of the surface waters. This process creates a zone of nutrient-rich water off the coasts of Chile and Peru, known for their high productivity and abundant and complex populations of marine life. When these equatorward winds fail, as periodically occurs, the upwelling ceases, marine productivity drops, and the fisheries dependent on the abundant phytoplankton fail. This periodic failure of the fisheries was the first observed symptom of the phenomenon we now know as the El Niño cycle, which varies over 3- to 7-year periods.

Another important example of seasonal variation in mixed-layer depth occurs in the North Atlantic, producing a phenomenon known as the spring bloom. During the winter, the North Atlantic is stormy, with high winds that cause mixing, which breaks down the vertical stratification of waters there. During the winter season, nutrient-rich water is mixed into the surface, and in the spring, warmer temperatures and increased sunlight lead to rapid, exponential growth of phytoplankton. Although the spring bloom is often portrayed as a wave, traveling north with sunlight and temperature, satellite observations show it to be a complex and patchy phenomenon, reflecting the chaotic influence of wind-driven mixing, changes to the temperature structure of the water column, and other intermediate- (meso-) scale (tens to hundreds of kilometers) phenomena. The bloom ends partially in response to the cessation of mixing. As the North Atlantic warms, it stratifies in the less-windy summer season, locking the nutrient-rich waters at depth. The abundant phytoplankton soon use all the available nutrients, and production slows. Just s important, as phytoplankton begin to grow, so do populations of herbivorous grazers (zooplankton) and higher trophic level organisms. As the populations of herbivores and carnivores increase, they begin to reduce the phytoplankton populations, contributing to the end of the spring bloom.

The North Atlantic spring bloom and the El Niño cycle result from coupling of climate to the oceans on relatively short timescales. The spring bloom is a seasonal phenomenon, while the El Niño varies over interannual timescales. However, the ocean varies on longer timescales as well. We know that the glacial–interglacial cycles are accompanied by about an 80 parts per million by volume (ppmv) change in atmospheric CO_2. This value is significant, representing a change in the atmospheric inventory of about 40%. These variations can't be due to the terrestrial biosphere, because the glacial world was cold and dry, and much of the currently vegetated land surface was covered by ice. The land actually stored significantly less carbon, but to explain the glacial cycle, it would have had to store more (Archer 2010). As Archer explains in detail, both paleorecords (stable isotopes) and models support this conclusion. Thus, the glacial–interglacial carbon cycles must be largely due to the oceans, but how?

Awkwardly, none of the known changes to the glacial ocean explain fully how the ocean was able to maintain a lower CO_2 concentration than at present. Many of the contributing causes are due to the physics of the oceans, such as the higher solubility of CO_2 in colder water, and these mechanisms are described in Archer's companion volume. Many of the changes to the glacial ocean, though, suggest very different ecosystems and accompanying changes to marine ecosystems. One provocative hypothesis coupled climate, dustiness, the nitrogen cycle, and marine ecosystems.

For many years, oceanographers have been mystified by so-called high-nutrient, low-chlorophyll (HNLC) regions in the oceans (Martin and Fitzwater 1988). These are areas in the open ocean where nutrients are sufficient to support high levels of photosynthesis (chlorophyll), but the chlorophyll levels remain low despite abundant nutrients. In the 1930s, oceanographers speculated that these areas might be iron limited, and in the 1980s, measurement and experimental techniques had improved sufficiently to enable this hypothesis to be tested. The HNLC waters were found to be extremely low in iron, and marine organisms require iron for key metabolic pathways, especially in the assimilation of nitrogen. Since iron is required for nitrogen uptake, and nitrogen is required for growth, this argument suggested that the HNLC regions were nutrient limited. What is the source of iron in these remote regions of the ocean?

Iron is not very available in oxygen-rich conditions, where it precipitates in chemically unavailable forms, so available iron does not persist in the open water column very long. However, dust transported to the oceans from terrestrial ecosystems can provide a source of biologically available iron. Most coastal regions, or areas with strong upwelling of deep water have sources of iron, either from dust or from chemical activity in low-oxygen deep waters. Interactions between decomposing organic matter and iron can also maintain higher iron availability in surface waters. The remote HNLC regions receive much less dust. Might this situation have been different in a glacial world? Many paleorecords—both direct measures of dust in ice cores and other chemical indicators—suggest that the glacial world was a dusty, dry, and windy place, with far more transport of dust to the oceans than in our interglacial world. The greater abundance of dust probably made glacial marine ecosystems significantly different from those today, with different patterns of productivity across the ocean basins, different dominant phytoplankton species, and altered biological feedbacks to atmospheric CO_2 and hence climate. Unfortunately, these effects are not sufficient to explain the 80 ppmv difference in atmospheric CO_2 between the glacial and interglacial periods, though they certainly contributed (Archer 2010). Current studies are focusing on the dynamics of the Southern Ocean.

In marine systems, the interactive roles of physical limitations, such as light and temperature, with nutrients have long been a core part of theory (Redfield 1934). In terrestrial ecology, two almost contradictory approaches to understanding ecosystems have evolved as a result of two seemingly contradictory patterns in nature. The productivity of ecosystems globally seems almost perfectly related to physical limiting factors. Most basically, the warm tropics are more productive than the cold poles, and wet forests are more productive than dry deserts. This relationship was first quantified by Michael Rosenzweig (1968), who combined temperature and moisture effects by plotting productivity against evapotranspiration (ET). Evapotranspiration combines water limitation (high ET requires an adequate supply of water) and energy limitation (high ET requires an adequate supply of energy to vaporize water). ET is also mechanistically correlated with photosynthesis via the stomatal control of photosynthesis, as discussed earlier. The response of plants to temperature, water, and sunlight is a part of everyday experience, and there is little doubt that these resources affect plant growth.

The observed relationship between ET and NPP has led to a wide range of theories to explain the ensuing pattern. Since light is the source of energy for photosynthesis, many models represent carbon uptake as a function of quantum use efficiency, that is, the amount of photosynthesis per unit solar radiation intercepted:

$$A = \varepsilon(\text{APAR})$$

where A is photosynthesis, ε is carbon fixed per unit light absorbed, and APAR is the amount of photosynthetically active radiation absorbed. These models are sometimes referred to as *epsilon* models after the symbol used to represent quantum use efficiency. They can be implemented for both terrestrial and marine ecosystems, but we will illustrate the approach for land. As written, this model implies that photosynthesis depends only on light absorption. From the earlier sections we know that this assumption is incorrect, and so the epsilon formulation is often combined conceptually with the Rosenzweig-type model by expanding ε to

$$\varepsilon = \varepsilon_f F(\text{H}_2\text{O}, \ T, \ L)$$

In this formulation, light-use efficiency (LUE) has a maximum, or fundamental, value (ε_f) that is reduced when water (H_2O), temperature (T), or other limiting factors (L) do not allow growth at maximal rates. This model seems simple, but it has a hidden but beautiful built-in complexity. The model describes an instantaneous rate of photosynthesis, A, as a function of absorbed photosynthetically active radiation and is conditional on soil moisture, temperature, and perhaps other factors. Is that sufficient to solve the equation? Actually, the PAR term is critical because it is defined as the amount (sometimes the fraction) of incident PAR absorbed by the vegetation. What determines the absorption of PAR? This absorption occurs in the leaves of vegetation (or other equivalent organs) and depends on the amount of foliage displayed, so APAR depends first on the area of foliage or leaf area index (LAI, defined as square meters of foliage per square meter of ground). Secondarily, APAR depends on the absorption per unit area of foliage, or the *optical properties* of the foliage, which are determined largely by their chlorophyll content and cellular structure.

The physics and biogeochemistry of terrestrial ecology are joined here, because for a plant to create leaf tissue and synthesize chlorophyll and other key biochemical, it needs nutrients, which occur in stoichiometric ratios in plants, just as in phytoplankton. Thus, although photosynthetic rate is determined by light, water, temperature, and other physical variables, the amount of light a plant can harvest to fuel photosynthesis may be limited by its ability to produce leaves. The LAI, in turn, may be limited by the availability of nutrients, just as in the oceans! The quantum yield equation has several parts:

$$A = \varepsilon_f(H_2O, \ T, ...) \times LAI_c(IPAR)$$

where IPAR is incident PAR, and LAI_c denotes LAI at a specific chlorophyll density, which determines its optical properties. LAI_c (IPAR) determines APAR. Nutrients affect both the LAI itself and its optical properties. In marine systems, the equivalent to leaf area is the density of phytoplankton and the abundance of their chlorophyll.

Very similar models can be written based on estimates of the absorption of light in the water column.

Net primary productivity, or NPP, is not determined solely by A but is actually A minus autotrophic respiration (R_a, or plant respiration). Some of the carbon fixed in photosynthesis is respired to fuel the plant's metabolism, and the amount of plant growth in a year is limited by the value of $A - R_a$. The epsilon model may be further expanded to

$$\text{NPP} = [\varepsilon_f(\text{H}_2\text{O}, T) \times \text{LAI}(\text{IPAR})] - R_a$$

The term R_a may be further expanded to encompass environmental controls over plant respiration rate.

More complex models recognize these additional controls over photosynthesis. In fact, carbon uptake by plants reflects not only light-use efficiency but the metabolic costs of growth, repair, and reproduction. The simple formulation of photosynthesis and productivity is robust but fails to predict many details of plant growth and will not necessarily work to predict changes under conditions different from those used to estimate LUE observationally. As all the limiting resources to growth change, the optimal plant response may result in changes to the LUE. Many of these interactions, and the link of photosynthetic rate to its underlying biochemistry, are captured in the *Farquhar model*, developed by Graeme Farquhar, Susanne von Caemmerer, and Joe Berry (1980). The full derivation of the Farquhar model is outside the scope of this book, but the paper by Farquhar, Caemmerer and Berry is one of the most important in Earth System Science and is the foundation for linking climate, nutrients, and carbon uptake by the biosphere. In the Farquhar model, photosynthetic rate is described by

$$A = V_c - 0.5V_o - R_d$$

where A is net photosynthesis, V_c is the rate of carboxylation (conversion of CO_2 to organic matter), V_o is the rate of oxygenation (formation of O_2 from H_2O) because of competition between oxygen and carbon dioxide in the photosynthetic enzyme system, and R_d is mitochondrial respiration. In this model, two key parameters emerge as critical controls over photosynthetic rate. The first is V_c (max), or the maximum rate of carboxylation, which appears as the critical rate control in the expanded equation for photosynthesis:

$$W_c = V_c(\max)\frac{C - \kappa}{C + K_c(1 + O/K_o)} - R_d$$

where W_c is the enzyme-saturated rate of photosynthesis; C is the intracellular partial pressure of CO_2; K_c and K_o are the Michaelis-Menten coefficients for CO_2 and oxygen, respectively; O is the intracellular concentration of oxygen; and κ is the compensation point for CO_2, that is, the CO_2 concentration where A is zero. The second critical parameter determining photosynthesis is $J_{\max}$, the maximum rate of carboxylation allowed by electron transport in the photosynthetic system. Photosynthesis occurs at either the enzyme-saturated rate (given in the preceding equation) or the enzyme-limited rate:

$$J = \frac{J(\max)}{2(2 + 2\phi)}$$

where ϕ is the ratio of oxygenation to carboxylation. The actual carboxylation rate is then computed as

$$V_c = \min(W_c, J)$$

In the Farquhar model, there are direct links to the physical world (all these equations can be expanded to show temperature and light dependence), but V_c (max) depends on the amount of photosynthetic enzyme, ribulose-1,5-bisphosphate carboxylase oxygenase (rubisco) and its ratio to chlorophyll. The abundance of this enzyme, in turn, depends on nitrogen, since it contains a great deal of nitrogen; most of the nitrogen in foliage is contained in the enzyme. Thus, at a biochemical level, this model captures the three-way interaction among the physical, the biogeochemical, and the biological components of carbon gain by the biosphere: when coupled to a model of transpiration, it illustrates how carbon gain and water use are coupled.

Models of carbon uptake have been integrated with the economic analogy by Rama Nemani and Steve Running (1989) to produce the *hydrological equilibrium* model. Nemani and Running observed tight correlations between LAI and water balance variables (precipitation, evapotranspiration) and developed a hypothesis to explain this observed pattern. They argued that plants would develop, on average, sufficient leaf area to use all the plant-available water at a given site because they would continue to develop leaf area until they became water limited. They would not add any additional leaf area beyond the LAI sufficient to transpire all the root zone moisture, because those additional leaves would constitute a respiratory demand (remember the economic analogy), but would not add any additional photosynthesis (since the site would have already run out of water). This prediction is widely satisfied, especially in drier ecosystems. Based on the preceding argument, this assumes that all the systems at hydrological equilibrium have sufficient nitrogen and other nutrients available to create sufficient LAI to use all the available water. This type of model suggests calculating NPP from *water-use efficiency* (WUE, carbon uptake per unit water) as well as light-use efficiency, and indeed, the epsilon-type model lends itself to this expansion.

WATER–ENERGY–NUTRIENT INTERACTIONS

All these terrestrial models based on physical limitations must confront an additional set of experimental constraints. Most terrestrial ecosystems, especially extratropical ones, respond to nutrient fertilization by increasing productivity. Recall that earlier we assumed that for the biophysical models to work, we had to presume that sufficient nutrients would be available that light or water and light- or water-use efficiency (LUE, WUE) would determine carbon gain. The positive responses of NPP to nutrients suggest that this assumption is not met and that many ecosystems are nutrient limited, as with the paradigm for the oceans. Countless papers show positive responses to added nutrients, and every farmer knows that crops become nutrient limited and require fertilization. In many ecosystems, nitrogen cycling rates are tightly correlated with productivity, suggesting that the natural rate at which nutrients become available in soils determines NPP.

Water/energy and nutrient control limitations are both consistent with our scientific knowledge, and with our everyday experience. First, plant growth is correlated with light, water, and temperature. Second, plants grow more when we add nutrient fertilizers. Are global patterns of productivity on land limited by water and energy or by nutrients? Are these patterns controlled similarly to or fundamentally differently from those in the oceans?

Are water/energy and nutrient controls over productivity consistent with each other, as in marine ecology, or are they alternative hypotheses, as many ecologists have assumed? One key to the puzzle is the difference in regulation of nutrient budgets between terrestrial and marine ecosystems. In marine systems, nutrients are input primarily through upwelling, although biological nitrogen fixation can be important in certain conditions, and most nutrients are lost through sinking of living and dead organic matter out of the euphotic zone. Nutrients tend to upwell in approximate Redfield stoichiometry, because they are derived from recycled dead organic matter, and of course, sinking organic particles tend to have nutrients bound to carbon in Redfield stoichiometry. The ultimate natural source of nitrogen for most terrestrial systems is biological nitrogen fixation (BNF), which transforms atmospheric N_2 into NH_4^+, a process that requires significant energy. Locally, fixed nitrogen may be redistributed via a wide range of processes. In the modern world, although biological nitrogen fixation remains important, on an annual basis industrial fertilizer production, inadvertent fixation of nitrogen in combustion processes, and crop fixation exceed natural biological fixation, and so locally, human sources of nitrogen may exceed biological ones.

Because nitrogen fixation, like photosynthesis, requires abundant energy, it too should increase with the abundance of energy in the environment. Empirically, rates of nitrogen fixation are observed to increase with increasing evapotranspiration. This is probably not because of a direct link between BNF and ET but because, as we've seen, high-ET systems can sustain high rates of carbon fixation (having both abundant water to support ET and adequate light and heat, plants can photosynthesize at high rates). When plants can fix abundant carbon, more energy is available to support BNF. Completing the cycle, at high BNF, adequate nitrogen is available to support continued high photosynthesis and organic matter formation (see figure 8).

Most nitrogen is lost either by solution as nitrate (NO_3^-) or by conversion of inorganic soil nitrogen back to atmospheric gaseous forms (denitrification), again mediated by the microbiota and requiring energy. Some gaseous nitrogen is also lost as ammonia (NH_3). Both of the key transformations (BNF and denitrification) involve fluxes of inorganic nutrients, require abundant

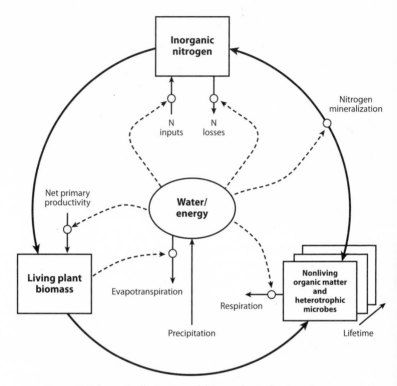

Figure 8. Schematic illustration of the coupling of water, nitrogen cycling, and carbon in ecosystems. Principle features of these coupled controls are that (i) water controls the inputs and outputs of nitrogen; (ii) increasing net primary productivity (NPP) allows more of the N flux through the system to be captured in organic matter; (iii) increasing organic N stocks allow for more N mineralization, which supports more NPP; and (iv) increasing precipitation allows both more NPP and more N cycling. Thus, water and nutrient limitation of NPP tend to become correlated.

(Source: Schimel, Braswell, and Parton 1997)

energy as carbon, need reducing conditions, and are favored by wet conditions, which allow high rates of microbial activity. Nitrogen losses increase in wetter climates. Wetter conditions allow more hydrological export, or leaching. Warmer conditions also enhance the activity of nitrifying and denitrifying bacteria, and increase nitrogen losses to the atmosphere as well.

These losses offset gains from BNF, but the balance is modified by the ability of ecosystems to form and store organic matter. Nitrogen losses are reduced when plants compete for inorganic nitrogen and fix it into stable organic forms. When plant demand is high, plants compete for mineral nitrogen and remove some of it from vulnerability to loss by fixing it into plant tissue. If plant growth is high, then the soil microbiota also compete for mineral nitrogen, needed for their growth also, which results in microbially stabilized organic nitrogen. Thus, as water and energy availability increases, nitrogen and carbon inputs increase, leading to high productivity. This cycle does not continue to indefinitely increase productivity because nitrogen losses also increase. Losses may increase dramatically if productivity reaches levels where all the available light is intercepted and/or all the available water is being used for photosynthesis, since the demand for nitrogen in new plant tissue will decline.

To generalize, the geography of nutrient limitation in the oceans is driven by the winds and ocean circulation, while the geography of nutrient limitation in terrestrial systems is controlled by geology and microbiology.

Nutrient inputs to terrestrial systems (BNF) are high where ample reducing power (organic carbon) is present, so to a first approximation, where NPP is high, BNF will be as well (Cleveland et al. 1999). Nitrogen inputs thus will be correlated with primary production and so should not produce a different biogeographic pattern than do water and energy. In fact, empirical rates of BNF are also correlated to ET. However, rates of nutrient loss are also linked to water and energy. Denitrification also requires reducing power (organic matter) and generally moist conditions to allow for the reduced, low-oxygen microsites where the denitrification reaction occurs:

$$2NO_3^- + 10e^- + 12H^+ \rightarrow N_2 + 6H_2O$$

As you can see, this reaction requires a large source of electrons and hence energy. Nitrogen losses are also likely to be high where energy is abundant and where wet conditions prevail, again paralleling NPP. What is the mechanism? In productive environments with warm, moist conditions, nutrient inputs are high, allowing for high production of biomass (and for high LAI, consistent with the epsilon model). Losses of N may also be high, but offsetting them, high production of biomass, with its stoichiometric ratios of carbon to essential nutrients, including N, forms a reservoir of N that allows the system to function at high productivity, with both high inputs and losses. Less productive systems have lower N fluxes but lower N requirements. Although disturbances such as droughts or wet periods, or episodic nutrient losses (such as wildfire), can disrupt this situation, where systems are more or less stable, water/energy and nutrient limitations should be correlated.

Some terrestrial systems are not nitrogen limited. Moist tropical forests are often more limited by phosphorus. Tropical forests have abundant energy available, with warm conditions and long growing seasons, and so ET and nitrogen availability can easily achieve equilibrium. However, at this point other nutrients, especially phosphorus, that do not have biological cycles but must be derived from local rock (or dust) can become limiting. This creates a situation similar to that in the HNLC regions of the oceans where nitrogen may circulate in abundance and not be tightly conserved within an organic cycle, because the ability of organisms to fully utilize the nitrogen is restricted by phosphorus. In such conditions, plants may adjust their Redfield ratios to increase their phosphorus-use efficiency (units of phosphorus per unit biomass).

In marine systems, climate can trigger ecological responses—often through nutrient supply—by altering mixing, salinity, and micronutrient supply (dust). In terrestrial systems, climate can also affect ecology through the nutrient cycles, but by affecting the rates of biological nutrient transformations. In natural terrestrial ecosystems, a major source of nutrients for plant growth is the decomposition of soil organic matter. As the organic matter is converted back to CO_2 (or simpler organic forms), nutrients bound in the organic matter are often released into plant-available forms. The decomposition processes are often accelerated by warmer or wetter conditions, and so if climate changes, nutrient availability will change, and water/energy and nutrient limitations may be out of equilibrium for some period of time, resulting in an increase in NPP, at least for a time. As described earlier, warmer conditions may also accelerate nutrient losses, so while warmer conditions may cause initial increases in productivity, the long-term consequences are harder to predict quantitatively.

Although patterns of climate and patterns of ecosystem structure and function are highly correlated, the underlying reasons for these correlations are driven by the fluxes of water and energy through living systems. These fluxes influence the availability of environmental resources (water, nutrients, and energy) that control carbon fixation and define the biological and physical resources for plants, animals, and microorganisms. They involve the adjustment of vegetation to the physical environment, microbial fixation, and loss of nitrogen, (and other nutrients) as well as trophic processes. While these processes differ greatly between terrestrial and marine environments, they also have a great deal in common. These energy and matter fluxes shape the response of species. The ways in which the evolved characteristics of species feed back to physical and biological phenomena are just beginning to be quantified.

4 ECOSYSTEM FEEDBACKS AND INTERACTIONS WITH CLIMATE

....

ECOSYSTEM EFFECTS ON CLIMATE

THE ECOLOGICAL AND CLIMATE SYSTEMS ARE INTER-connected by many feedbacks and interactions. Ecological systems affect climate by influencing the atmospheric concentrations of greenhouse gases, aerosols, and water vapor. Ecosystems directly affect the surface energy budget through albedo and the partitioning of energy between forms (sensible and latent heat). The climate system, in turn, affects ecosystems through the multiple variables described in chapter 2, coupling physical and living systems together.

The carbon cycle is central to critical Earth System feedbacks (Archer 2010; Doney and Schimel 2007). Terrestrial and marine carbon budgets are sensitive to climate, and the climate system is very sensitive to atmospheric CO_2. Climate changes triggered outside the carbon cycle may be damped or amplified by carbon cycle feedbacks, leading to complicated interactions between orbital and geological forcing of climate and the carbon cycle (Archer 2010). Carbon–climate coupling is nowhere more evident than in the record of CO_2 and temperature

....

from two critical environmental records, the Keeling record and the long ice cores from Antarctica and Greenland. These two records frame carbon–climate coupling
in time and reveal how CO_2 and climate have varied together over months, years, and millennia. As carbon science has developed, extraordinary observational records
collected by a series of exceptional observational scientists have inspired questions and informed their answers.

Beginning in 1958, Dave Keeling began measuring CO_2
at the summit of Mauna Loa, in Hawai'i. The development
and maintenance of the measurements to the present day
is a remarkable story, to which this text can't do justice
(but see Keeling 1998). Even a quick look at the Keeling
record (see figure 9) gives some fundamental information

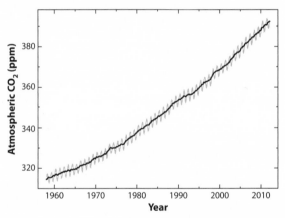

Figure 9. Atmospheric CO_2 at Mauna Loa Observatory. The Keeling
curve, showing the steady increase in atmospheric CO_2, the biologically driven seasonal cycle, and interannual variations in growth rate
(see figure 10).

(Source: Scripps Institute of Oceanography and NOAA 2012)

about the modern carbon cycle. First, the level of CO_2 is rising, in parallel with increasing human emissions of carbon from fossil fuels and changing land use, or deforestation. When Dave Keeling began this record, no one knew what would happen to fossil fuel CO_2: Would it all accumulate in the atmosphere? Would all the CO_2 be absorbed by the oceans? Keeling's data showed that fossil carbon did accumulate in the atmosphere. Radiocarbon measurements soon showed the carbon accumulating was indeed the ancient radiocarbon-dead carbon from fossil fuels, as determined from the change in the ratio of atmospheric concentration of the isotopes of carbon known as the *Suess effect*, after Hans Suess who first discovered it. When scientists first began to worry about fossil fuel carbon, the fate of this carbon in the Earth System was not known.

After it was discovered that fossil fuel carbon accumulated in the atmosphere, it soon became apparent that not all the fossil fuel emissions were accumulating there. Keeling was measuring the change in atmospheric CO_2, and he knew approximately how much fossil fuel was being burned from energy statistics, so he very quickly realized that *significant amounts of carbon were being absorbed by the Earth System*, that is, by the oceans or biosphere. The effect of fossil fuel burning on climate depends on the resulting atmospheric concentration of CO_2, since the amount of heat trapped depends on the concentration (recall chapter 1). If Earth System processes affect the relationship between *emissions* of CO_2 and the resulting atmospheric *concentrations*, then those processes will modulate the climate effect of fossil fuels. If climate affects these Earth System processes, *complex carbon–climate feedbacks can ensue* (Woodwell et al. 1995).

However, the Keeling record informs more than these three momentous discoveries (of the increase in CO_2, its uptake by the Earth System, and the potential for carbon–climate feedbacks). One of the notable features of the curve in figure 9 is the strong seasonality it shows, as an oscillation around the upward trend. This oscillation, called the *breathing of the Earth* results from Northern Hemisphere summer (where most of the planet's land lies) photosynthesis, which draws down atmospheric CO_2 and respiration, or the winter decomposition of dead plant material. This was one of the least-anticipated discoveries, because when Keeling began his work, climate science was dominated by geophysics, and no one suspected that life could so visibly and systematically alter the atmosphere. The seasonal cycle provides global quantification of the effect of seasonal climate on ecosystems. The global atmosphere provides us with a data set describing the effect of climate on net carbon uptake and storage by plants and soils globally, replicated every year, if we are smart enough to understand how to relate it to measurements of leaves and individual plants at scales vastly finer than those of Keeling's data.

Still, with this fourth discovery, of the breathing of the Earth, scientists had just begun to mine this unique data set. A close examination of the Keeling record reveals a number of slower *wiggles* where the steady rise in CO_2 seems interrupted for a short while. These carbon cycle wiggles record an astonishing range of Earth System events. Let's look at the period after 1973. In 1973 the Arab–Israeli war was followed by an oil embargo, which led to a massive spike in fuel prices, reduced consumption, and adoption of more efficient practices, worldwide. The growth rate of CO_2 in the atmosphere reflects that brief reduction in the upward trend in energy consumption (see figure 10).

Twenty years later, we see another flattening of the growth rate, around 1993. In this case, the growth rate was not due to a geopolitical event but resulted from the eruption of Mount Pinatubo, in the Philippines. The exact mechanism for this slowdown is debated, but for a year or more, terrestrial carbon storage increased. The

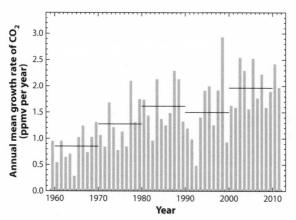

Figure 10. Annual mean growth rate of CO_2 at Mauna Loa Observatory. Year- to-year changes in the growth rate of atmospheric CO_2 reflect Earth System events (El Niño years, volcanoes) and human history (the oil embargo, collapse of the Soviet Union).

(Source: Scripps Institute of Oceanography and NOAA 2012)

Pinatubo event caused cool temperatures, which may have reduced respiration. The volcanic aerosols also caused a global increase in diffuse radiation, which may have increased photosynthesis. As the aerosols were scrubbed from the atmosphere, there was a decrease in the effects, and the growth rate returned to normal.

In 1997, the growth rate of CO_2 spiked high, doubling the normal rate. This phenomenon implied that terrestrial uptake dropped to zero for that year, but why? In fact, terrestrial uptake may have been normal that year, but it was an El Niño year, and the tropical droughts that frequently accompany El Niño were severe. In Indonesia, there is a region of tropical peatland forests, in Kalimantan province. Those forests are being harvested, and to harvest them, the loggers cut canals back into the woods and lower the water table. The extreme El Niño droughts of 1997 triggered wildfires in Kalimantan, and vast amounts of stored carbon were lost to the atmosphere, causing much of the increase in atmospheric CO_2. El Niños are frequently accompanied by carbon-cycle disturbances, and they are often due to carbon release from wildfires associated with drought.

In some El Niño years, the growth rate of CO_2 drops early in the El Niño, so in 1997, there was a slight flattening before the steep rise in CO_2. El Niño is also an oceanic phenomenon. Normally, the global ocean circulation of water transfers high-CO_2 water from high to low latitudes, and this water upwells in the tropics. During El Niño, the tropical winds change direction, and warm water caps the upwelling regions. This temporarily traps the cold, high-CO_2 water at depth and prevents it from entering the atmosphere, leading to the slight flattening of atmospheric CO_2 early in El Niño periods.

From the one seemingly simple, measurement of CO_2, we have learned a remarkable amount about interactions of the carbon and climate systems. But, wait, there's more! Soon after starting measurements at Mauna Loa, Keeling and others began to make measurements elsewhere around the world, often in remote and challenging locations, and they noticed another regularity. Concentrations of CO_2 were higher in the Northern Hemisphere than in the Southern Hemisphere. This made sense, since the bulk of fossil fuel emissions occur in the Northern Hemisphere. By this time, enormous amounts of data on oceanic CO_2 had been collected, and so, not only could ocean carbon uptake be estimated, but maps of where the ocean absorbed and released CO_2 could be produced (Takahashi et al. 1998). Combining these maps with the atmospheric gradient of CO_2 in a model, Inez Fung, Taro Takahashi, and Pieter Tans (Tans, Fung, and Takahashi 1990) found that the model predicted an excess of CO_2 in the Northern Hemisphere atmosphere compared with observations. Some real-world process, not included in the model, was taking up carbon in the Northern Hemisphere, and constrained by Takahashi's data on the oceans, the analysis suggested it had to be on land.

Tans et al. identified, based on atmospheric data, a Northern Hemisphere, probably midlatitude, uptake of carbon. They made the identification basically by solving a mass balance equation for the atmosphere, so that the change in atmospheric concentrations over time was a function of known inputs (fossil fuels), minus known removals (the ocean), plus an unknown additional removal (sometimes called the *missing sink*). The spatial distribution of the inputs and removals was constrained by the observations of atmospheric concentrations latitudinally and by Takahashi's map of ocean CO_2 uptake. This was an enormously important result, but it had even bigger implications for biology.

Why was this result so important? From simple calculations, by the time of the Tans et al. paper it was clear that ocean uptake could not account for the discrepancy in the carbon budget that Keeling had identified: that is, the oceans could not explain the difference between fossil fuel emissions and atmospheric concentrations. The prevailing hypothesis involved uptake of carbon by terrestrial ecosystem: as CO_2 increases, plant photosynthesis becomes more efficient. Theory and countless laboratory experiments had shown that increasing CO_2 could lead to enhanced terrestrial uptake. Models of this mechanism suggested the effect should be more or less proportional to NPP, so the effect should be large in the tropics and productive forests, and small at high latitudes (Mooney et al. 1991). Theory also predicted the effect could be large in dryland ecosystems, but probably not to global ecosystem carbon storage, as dryland systems tend not to contribute large fluxes to the carbon cycle. However, Tans, Fung, and Takahashi found an effect with a very different spatial pattern. They suggested uptake by midlatitude ecosystems and not by tropical ecosystems. If this was true, was CO_2 *fertilization* the mechanism or not? We will revisit this question later.

The Tans et al. paper was a landmark in Earth System science. A study based on data from remote locations far from ecosystems, analyzed using a physical model of transport in the atmosphere by the winds, and constrained by chemical observations in the oceans derived a conclusion of fundamental importance for terrestrial biology. To understand how the global terrestrial biosphere functions, researchers have to understand atmospheric circulation, marine physics, and biogeochemistry and have hypotheses about the biology derived from the discipline of ecology. This grand unification of the environmental sciences is sometimes termed Earth System Science, so named by Francis Bretherton from the National Center for Atmospheric Research in the United States. Many problems of ecosystems and climate can be understood only by taking this approach.

The Keeling record now spans 53 years and records the influence of climate on ecosystems through the seasonal cycle, the El Niño cycle, and other short-term patterns. If there is an effect of gradual warming on the average uptake of carbon by the biosphere and oceans, it remains subtle. However, another record of CO_2 collected over a similar period of time, beginning in the 1950s, spans a much longer period of time. In the 1950s and subsequently, glaciologists led by Willi Dansgaard realized that ice in the polar ice caps formed in layers and preserved environmental records. The first record retrieved studied oxygen isotope ratios, which preserve a record of temperature, because when water changes phases, the abundance of the heavy and light (^{18}O and ^{16}O) isotopes changes in proportion to temperature.

Soon after, Hans Oescher realized that the ice layers might also preserve CO_2 concentrations from the past. As ice forms in the Greenland and polar ice caps, air permeates the *firn*, or partially compacted snow. Eventually, under the right conditions, such as in Antarctica, the firn forms ice, and the air remaining is trapped in bubbles that preserve the air's chemical and isotopic composition. The *Vostok* core is a record of the last four glacial cycles (420,000 years) and shows how temperature, CO_2, and CH_4 (and other markers) varied as the glaciers waxed and waned.

These records are long enough to show the effects of climate on carbon, as well as of carbon on climate (see figure 11). Although there seems to be a simple correlation between the temperature records and the CO_2 concentrations, the actual interaction is very complicated. The ice age cycle is triggered by the slight instability of Earth's orbit, which leads to subtle changes in the amount of sunlight the Northern Hemisphere receives over time. These changes in insolation trigger the expansion of ice sheets, which creates a reinforcing feedback. The ice sheets are bright—brighter than the open water, soil, and vegetation they replace—and reflect sunlight back to space efficiently, which reinforces the cooling due to the orbital changes in sunlight and allows the ice sheets to expand. Changes in atmospheric CO_2, CH_4, and N_2O parallel the changes in temperature. Do changes to

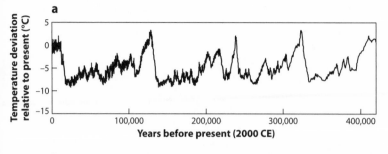

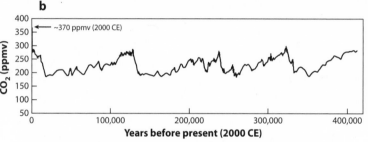

Figure 11. (a) Temperature of lower atmosphere over the last 400,000 years, from Antarctic ice and air data. (b) Atmospheric CO_2 concentrations over the last 400,000 years, from Antarctica ice and air data. The glacial–interglacial record of temperature and CO_2, from the 420,000-year-long Vostok ice core, showing temperature–carbon cycle correlations across four glacial cycles.

(Source: Petit et al. 1999)

temperature cause the changes in the carbon cycle, or vice versa? Very detailed analyses show that temperature changes before CO_2 and the other greenhouse gases affect it. This is just as expected, since the temperature changes are paced by changes to Earth's orbit. CO_2 produces about a third of the temperature change between the glacial and interglacial periods. The carbon cycle *reinforces and amplifies* the glacial–interglacial cycles; in other words, there is a feedback effect. Or, as Claude Lorius and colleagues wrote in 1990,

> changes in the CO_2 and CH_4 content have played a significant part in the glacial–interglacial climate changes by amplifying, together with the growth and decay of the Northern Hemisphere ice sheets, the relatively weak orbital forcing . . .

Analyzing the glacial–interglacial temperature and carbon cycles, we can estimate the strength of the carbon–climate feedback, which, in fact, corresponds to about an equilibrium 3° global increase in temperature for each doubling of CO_2, which is very much in line with estimates from other calculations.

Although estimating the warming effect of paleo-CO_2 changes provides useful information, when we evaluate the effect of modern-day CO_2 emission, we must also know how long the effect of a change in concentration lasts. Many gases (see methane discussion later) are oxidized into simpler compounds in the atmosphere, and so they have a well-defined *lifetime* over which they are irreversibly converted into some other form. The concentration of these simpler gases declines exponentially with time, as described by the following equation:

$$C(t) = C(0)e^{-t/\tau}$$

where $C(0)$ is the initial concentration, $C(t)$ is the concentration at some time t, and τ is the total lifetime of the substance.

Carbon dioxide is not oxidized in the atmosphere but may be absorbed by the oceans or taken up by the land. However, these processes are reversible, so once CO_2 is taken up, it may be returned to the atmosphere. When fossil fuels are burned, carbon is transferred from forms such as coal and oil, where it was stable for millions of years, and is returned to the active land–ocean–atmosphere cycle. Thus, changes to atmospheric CO_2 will persist at some level until the fossil carbon is once again in a stable and unreactive form.

The longer CO_2 remains in the land/atmosphere, the more it affects the climate. Even after a reduction in emissions, the climate effects persist. Thus understanding the lifetime of atmospheric CO_2 concentration changes is critical. Because CO_2 recirculates in the atmosphere–land–ocean system, its effect lasts until it is first stored in deep ocean reservoirs and finally sequestered in sparingly soluble mineral forms (see Archer 2010). This process takes thousands to tens of thousands of years. Because CO_2 recirculates, its disappearance does not follow the exponential law described in the preceding equation but instead shows an initial drop, followed by a very long, slow disappearance, so the climatic effect of fossil fuel emissions may last thousands of years. Most estimates of the initial half-life (the time required for the concentration of CO_2 to drop by 50%) are 50–200 years (depending on assumptions about the biosphere, changes in ocean circulation, and other factors; Moore and Braswell 1995), but it takes 10,000 or more years for the concentrations to return to preindustrial levels after all emissions cease (an unrealistic situation for the foreseeable future). The remote paleorecord can be used to estimate this long disappearance time, although the preglacial world was sufficiently different that this is only an approximation (Doney and Schimel 2007).

The relationship between emissions of CO_2 and resulting concentrations in the atmosphere is complicated by carbon's complex pattern of natural release and uptake through many Earth System processes. When scientists and policy makers began to think about trying to manage fossil fuel burning to eventually stabilize the atmospheric concentrations, scientists realized that they did not know the relationship between emissions and atmospheric concentrations well enough to predict the impact of any given reduction in emissions on concentrations, and hence climate. In the mid-1990s, an international project was begun to understand these relationships, given all that was then known about the role of the land biosphere and oceans.

This project used a range of models of the carbon system that included what was known about ocean carbon uptake and terrestrial carbon storage, together with how those processes respond to changing atmospheric CO_2 (Enting, Wigley, and Heimann 1994). One early result was that stabilizing present-day emissions (emissions of fossil fuel are currently increasing each year) would not stabilize atmospheric concentrations but instead would lead to linear increases forever! Although models did not agree exactly, all showed similar patterns, indicating that stabilizing atmospheric concentrations at levels between

350 ppmv and as high as 1000 ppmv (more than three times the preindustrial level of 280 ppmv) would require reductions of 30%–90% from 1990 fossil fuel emission levels. That is, the emissions would have to decrease to levels low enough for Earth's natural carbon uptake and storage processes to sequester the added carbon long enough for it to be finally converted to stable organic or inorganic forms. These analyses were done optimistically, assuming that climate change would not itself cause further emissions or weaken land and ocean uptake processes. If ecosystem or ocean mechanisms were weakened, then reductions in emissions would have to be even bigger to allow the smaller Earth System uptake processes to keep up.

CARBON STOCKS AND FLUXES

Measuring CO_2 concentration changes in the atmosphere is a powerful way to study the carbon cycle. Although such changes can be related to processes occurring in the biosphere and oceans, these processes occur at the scale of molecules, cells, organisms, and biological communities. The atmosphere records the sum of all these microscopic processes, but it does not allow us to define precisely what processes occurred and where. To understand carbon cycle processes and feedback with climate, we also need to observe and experiment on the actual mechanisms. And to understand the mechanisms, we need to understand the Earth System components (atmosphere, oceans, and terrestrial biosphere) that interact with atmospheric CO_2

and the exchanges of carbon among them. Knowing the amounts, forms, and distribution of carbon in the land and ocean reservoirs, we can begin to understand how this carbon might interact with the climate system to produce changes in atmospheric CO_2.

Including those Earth System components that interact with climate over years to centuries, the atmosphere itself contains about 780 billion metric tons (or gigatons, Gt) C, equivalent to an average concentration of 392 ppmv (as of 2012) (see figure 12, reflecting slightly earlier data). The oceans contain the largest amount of active carbon, or about 38,000 Gt C. The earth's lithosphere (rock) contains even larger but mainly inactive reservoirs of carbon that

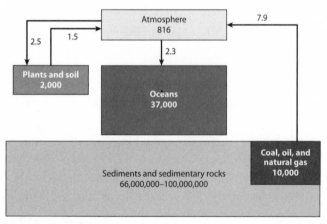

Figure 12. Global stocks and flows of carbon (in Gt C) in different Earth System reservoirs, and the fluxes among them, including the current emissions from fossil fuel and deforestation and uptake of fossil fuel/deforestation CO_2 by the land and oceans.
(Source: Global Carbon Project 2011)

cycle only on geological timescales (Archer 2010). Carbon in the deep ocean equilibrates with atmospheric CO_2 over about a thousand-year timescale. This equilibration time is long because, first, the amount of carbon involved is very large, and second, because the ocean is stratified, it takes 500–1000 years for surface water to mix with the deep water and thus to show the effect of CO_2 changes in the atmosphere on the deeper waters. The ocean inventory of carbon was estimated by combining hundreds of thousands of samples collected by oceanographers and producing maps from the data. Compared with the land biosphere, the ocean is relatively well mixed, so, especially in the deep ocean, a single measurement represents a fairly large area.

The terrestrial biosphere contains a substantial amount of carbon (Schimel 1995). Unlike in the Earth's great fluid media (the atmosphere and oceans), carbon in the land is not well mixed but is instead highly variable. Estimates of land carbon storage are built up laboriously by assembling measurements of vegetation and soils, mapping them, seeking regional common values, and then summing these regional numbers to form global estimates. Earth's vegetation contains about 500 Gt carbon, mainly in the tropical and boreal forests, because they have high carbon storage and cover large areas. Soils contain far more carbon than vegetation, or about 1500 Gt C. The patterns of soil carbon storage are not identical to those of vegetation. Most vegetation carbon occurs in forests, but grasslands and Arctic tundra soils contain huge amounts of carbon.

The Earth System contains about 40,000 Gt of actively circulating carbon, with about 90% of that in the oceans. Only 2% of the active carbon resides in the atmosphere. The increase in atmospheric CO_2 represents an even smaller fraction of the active pool, about 0.5%. However, Earth's carbon cycle is so finely balanced that that addition of 0.5% carbon to the atmosphere can trigger large changes that will last for thousands of years. The addition of fossil carbon to the active cycle causes large changes in the annual exchanges of carbon among its component reservoirs, accompanied by changes to atmospheric concentrations and hence climate.

What are the annual exchanges of carbon among the reservoirs? We know for example, from the Keeling record, that the biosphere both absorbs (photosynthesis) and releases (respiration and combustion) carbon over the annual cycle, causing the characteristic oscillation in the data. Thus, on an annual basis, the net exchange is the difference between uptake and release. In fact, the land biosphere takes up about 120 Gt/yr C and releases a similar amount back to the atmosphere. The oceans take up carbon in areas of sinking dense (salty), nutrient-rich waters and release carbon when these waters once again mix back to the surface. The ocean takes up about 90 Gt C and—much like the land biosphere—releases a similar amount back to the atmosphere. While the uptake and release of carbon by the biosphere occurs mainly in different seasons, uptake and release of carbon by the oceans occurs mainly in different regions: in both cases the annual signal seen by the atmosphere is only a few gigatons of carbon (see figure 12).

If the land and ocean fluxes were perfectly balanced, then all the carbon released by fossil fuel burning would accumulate in the atmosphere, but we know from the atmospheric observations and fossil fuel statistics that it does not. The atmosphere accumulated 4.1 Gt C/yr in the first decade of the twenty-first century, because the land and ocean fluxes were not perfectly balanced (although they might have been nearly so preindustrially). During that time period, fossil fuel burning released an average of 7.9 (±0.5) Gt C/yr. The oceans absorbed a net of 2.3 (±0.4) Gt C; that is, uptake exceeded release by 2.3 Gt.

The role of the land is more complex. In addition to the 7.9 Gt C/yr from fossil fuel burning, scientists estimated that land use, mainly deforestation, released an additional 1.1 (±0.7) Gt C/yr. To balance this budget, the land (in addition to losing 1.1 Gt C/yr) must also gain an additional 2.4 Gt C/yr. The following balance equation gives a sense of the entire budget:

$$(\text{inputs}) - (\text{uptake} + \text{atmospheric accumulation}) = 0.0$$

or

$$(7.9 + 1.1) - (4.1 + 2.3 + 2.5) = 0.0$$

Of these components, the estimates of deforestation and land uptake are the most uncertain, and the most variable with time. The uncertainty arises from the difficulty of making measurements of terrestrial carbon fluxes and stocks over large areas. Deforestation, for example, occurs as many small events within large regions. While the area deforested can increasingly be estimated using space-based observations, the actual amount of carbon lost to the atmosphere (or gained as vegetation recovers) has to be estimated with field studies. Similarly, carbon uptake by the biosphere varies with soil type, topography, species composition, and the history of the site (the length of time since a fire or other disturbance). Freshwater ecosystems (rivers and lakes) add an additional dimension to the coupled carbon cycle. Transport of dissolved and eroded organic matter through watersheds is very substantial, equal in magnitude to the annual uptake of fossil fuel carbon (~2 Gt C/yr) and so changes to the water cycle and erosion regimes can have significant effects on terrestrial carbon storage (Aufdenkampe et al. 2011). Inputs of carbon and alkaline material may also have significant effects on ocean biogeochemistry, especially in the coastal zone. The coupling of terrestrial, aquatic, and oceanic carbon cycles may cause unforeseen changes as hydrological cycle changes interact with changing land use patterns.

Because biospheric carbon exchange patterns reflect very local conditions (soils can vary greatly over meters), building budgets over millions of square kilometers requires vast amounts of data and innovative techniques for combining these data. Getting a sense of the average biomass and soil carbon in a region is hard, as is quantifying transport and storage in rivers and lakes. Estimating changes over time is even more difficult. Understanding why changes have occurred is yet harder still. Without knowing why observed changes to carbon storage have occurred in the past, it is difficult to project how carbon cycle processes will operate in the future and what feedbacks to climate might ensue.

Land uptake is variable over time. When land plants take up carbon, they discriminate against the heavy isotope of carbon, ^{13}C, whereas ocean uptake does not. Thus, the mass balance of ^{13}C records year-to-year changes in the balance between land and ocean uptake. From the ^{13}C tracer we know that most of the variability in the growth rate of CO_2 is due to the impact of climate on terrestrial ecosystems. The global record allows some climate–ecosystem feedbacks to be directly analyzed. For example, year-to-year changes due to biomass burning can be related to atmospheric changes by remote sensing and on-the-ground reporting of fires, so the impact of the climate (usually drought) on wildfire can be understood. Some climate–ecosystem interactions are more complex, such as the impact of the Pinatubo eruption described earlier. Quantification of the impact of climate on terrestrial carbon exchange with the atmosphere demonstrates that rates of carbon exchange respond to climate. To understand how climate affects the myriad organisms that are responsible for terrestrial carbon uptake and release, we need to use other approaches.

The most direct way of observing how local terrestrial carbon exchange responds to climate is by using a technique called *eddy covariance*, although many techniques have been used over the years. Eddy covariance measures the tiny variations in carbon dioxide in rising air (carrying the signal of terrestrial carbon uptake or release) and the velocity of the air motion (Baldocchi 2003) compared with sinking air (which has atmospheric concentrations). The biosphere either depletes (photosynthesis) or enriches (respiration) air relative to the atmosphere, and eddy covariance is able to quantify this effect before mixing dilutes the biospheric signal. Since the small turbulent eddies that carry this signal occur at high frequency (varying faster than a second), the eddy covariance measurement is necessarily made using high-frequency instruments. Changes in carbon exchange can be resolved over periods as short as half an hour. Integrating half-hourly measurements allows variations over synoptic variability (days to a few weeks), the seasonal cycle, and between years to be observed. Thus, direct effects of climate on carbon can be quantified over years. However, the eddy covariance method captures carbon exchange over only a square kilometer or two, at most. Since the climate sensitivity of ecosystem carbon storage varies with species composition and age of plants, as well as with site conditions such as soil water-holding capacity and nutrient content, how can local methods be used to explain global carbon cycle variations?

Eddy covariance sites are usually established by researchers with an interest in the carbon dynamics of a particular type of system, but most of the existing sites are organized into networks of collaborating researchers. These networks of sites allow regional patterns to be analyzed. One of the best examples is an analysis of the impact of the pan-European drought of 2003. In that year, Europe experienced extreme high temperatures and lower-than-average rainfall. Crop yields were amongst the lowest in a century. Comparing data across 14 widely distributed European sites, Philippe Ciais and his colleagues found dramatic reductions in photosynthesis and carbon uptake during the drought year, and strong evidence that reductions in soil moisture from both the low rainfall and the high evaporative rates were the cause. Although the 14 sites spanned much of Europe and showed results consistent with satellite observations of vegetation and reported crop yields, the global atmosphere also recorded the impact. The period 2003–2004 had one of the highest growth rates of CO_2 ever observed. Recall that reduced uptake of carbon by the land biosphere causes an *increase* in the growth rate of CO_2 in the atmosphere. Ciais and his colleagues estimated that Europe-wide, the continent might have contributed 0.5 Gt C to the atmosphere, increasing the global growth rate by 0.25 ppmv/yr.

Eddy covariance allows for carbon exchange processes to be observed directly in terrestrial ecosystems, mainly by tracking ecosystem responses to light, temperature, and water. By comparing sites, effects of species, tree age, and nutrient status may be analyzed, but rarely do these factors change fast enough on land for their effects to be observed directly. In marine ecosystems, photosynthesis and carbon storage can also be observed and the underlying mechanisms understood, but the approaches, timescales, and processes involved differ a great deal. Marine scientists, like terrestrial scientists, maintain networks of sites where carbon cycle processes are observed continuously. While eddy covariance requires expensive instrumentation, seagoing time series are even more costly and difficult to maintain.

The North Atlantic is one of the major sites for ocean uptake of carbon and may account for up to nearly a third of global ocean uptake. Climatically, the North Atlantic is affected by a well-known climate pattern known at the North Atlantic Oscillation (NAO), which causes coordinated changes over much of the basin (Wang and Schimel 2003). The NAO is caused by differential air pressures between a subtropical high near the Azores and an arctic low near Iceland. The NAO causes changes in North Atlantic water temperatures; positive NAO indexes are generally correlated with warmer temperatures and weaker winds, and negative NAO indexes are correlated with colder temperatures and stronger winds. The North Atlantic is one of the few regions with all the data needed to understand the complex relationships involving the physics, chemistry, and biology that govern the ocean carbon cycle. The Bermuda Institute of Ocean Science has been collecting data since 1988 in the Bermuda Atlantic Time Series study (BATS), and so detailed analyses of those local data can be used to develop a mechanistic understanding linked to observations of larger-scale patterns observed in the atmosphere (see figure 13).

Temperature directly affects the solubility of CO_2: warmer water absorbs less carbon dioxide. Warmer years

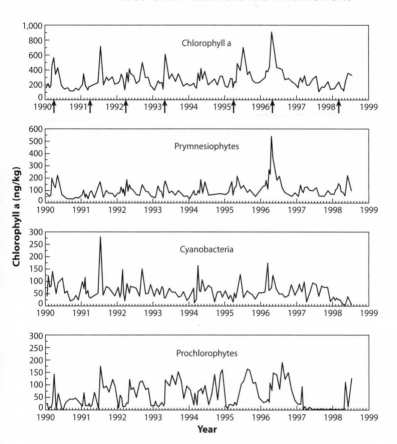

Figure 13. Chlorophyll (related to biological carbon uptake) and changing species composition over time at the Bermuda Atlantic Time Series site, showing the correlation among species changes, chlorophyll, and—ultimately—primary productivity. Prymne-siophytes, cyanobacteria and prochlorophytes are key functional groups of phytoplankton. Arrows indicate the timing of phyto-plankton blooms.

(Source: Steinberg et al. 2001)

have less mixing (typically with a positive NAO index), and so less dissolved CO_2 and fewer nutrients are mixed into the euphotic zone. As a result, the phytoplankton are less productive. There is less uptake of CO_2 from the atmosphere because of warmer sea surface temperatures and reduced wind speeds. Colder years exhibit deeper mixed layers with larger entrainment, enhanced net community productivity, and higher CO_2 uptake from the atmosphere. The variations in North Atlantic carbon exchange predicted from BATS data agree closely with some estimates of North Atlantic carbon exchange estimated from atmospheric CO_2 measurements. The BATS and global data, taken together, suggest that North Atlantic uptake varies by nearly 50% depending on the state of the NAO. Changes to the NAO are thought to be linked to climate change, so these studies suggest a warmer North Atlantic (similar to the positive NAO) may take up less carbon and so contribute to even warmer climates.

These measurements link climate, ocean circulation, nutrient supply, and biological productivity, and are similar to the types of analyses that can be done in terrestrial systems. However, because the oceans contain mainly microscopic organisms, species and populations can change very rapidly. By monitoring chemical pigments that are diagnostic of particular groups of organisms, BATS also charts the changes in plankton species composition within and between years (Steinberg et al. 2001). While changes in carbon dynamics can be empirically explained by temperature, wind, and nutrients, many of the changes in phytoplankton uptake of carbon and subsequent storage of this carbon in the deep ocean depend on the species present. When conditions favor high productivity, and nutrient availability is high, the fraction of carbon fixed by photosynthesis that is eventually stored in the deep ocean is likewise high. High productivity may also occur when conditions favor nitrogen-fixing cyanobacteria, which can support significant carbon uptake in the absence of nitrate from below. Conversely, unlike plants on land, which are fixed in location and long-lived, marine organisms change rapidly. Marine and terrestrial time series may be combined with global atmospheric data to explain observed patterns in the planetary atmosphere, but the types of mechanisms revealed are astonishingly diverse: marine observations reveal species changes within a year of a magnitude that would require decades to play out on land.

A potentially important aspect of the carbon–climate coupling is the response of land photosynthesis to increasing CO_2 (CO_2 fertilization), which was initially assumed to be an important aspect of the carbon cycle response of terrestrial ecosystems. From first principles, we know that, other things being equal (light, water, nutrients), increasing CO_2 should increase the efficiency of photosynthesis. The first reason lies in the enzyme kinetics of the photosynthetic enzymes. As carbon dioxide concentrations rise, the rate at which carbon is fixed into simple sugars increases until limited by other factors. The main enzyme that captures carbon dioxide, ribulose-1,5-bisphosphate carboxylase oxygenase (rubisco), binds both carbon dioxide and oxygen. When the concentration of carbon dioxide is high, rubisco fixes CO_2. However, when the carbon dioxide concentration is low, oxygen competes at the binding site and replaces CO_2. This process, *photorespiration*, uses energy but does not fix carbon. Thus, as the concentration of CO_2 relative to O_2 increases, the rate of carbon fixation increases.

There is a second effect of increasing CO_2. Photosynthesis occurs within the leaf, and CO_2 and water vapor diffuse in and out through the stomata. The rate of CO_2 into the leaf is governed by the gradient of CO_2 concentrations inside the leaf (c_i) relative to atmospheric CO_2 outside the leaf (c_a) and the leaf conductance (G_c), which is determined by the stomatal aperture. The driving force for diffusion is

$$(c_a - c_i)$$

This means that the higher atmospheric CO_2 rises, the larger the difference $(c_a - c_i)$ becomes, because c_i is determined by photosynthetic rate. For any given rate of CO_2 uptake, at higher $(c_a - c_i)$ the stomates need not open as wide. Recall that the wider the stomates open, the more CO_2 enters the leaf but the more water is lost. For any given rate of CO_2 uptake, at higher values of $(c_a - c_i)$ the stomates need not open as wide. This means that less water is lost per unit CO_2 entering the leaf, which increases the *water-use efficiency* (WUE) of the plant and could allow it to thrive under drier conditions. Not only does CO_2 fertilization affect the carbon cycle, but it alters the response of ecosystems to climate change, creating additional feedbacks. The relationship between increasing CO_2 and WUE has led to a hypothesis that the impact of increasing CO_2 would actually be larger in drier ecosystems or during dry periods.

The response of photosynthesis follows a familiar form, described by a Michaelis-Menten curve. Photosynthesis is zero at some positive value of CO_2 where photorespiration equals photosynthesis, and rises to a maximum where CO_2 is saturating and other reactions become rate limiting. Taking typical experimental values, we expect photosynthetic rates to increase by 25%–40% between the preindustrial concentration of CO_2 and double that value (280 ppmv versus ~550 ppmv). Although this is what we would expect if all other limitations were equal, is this effect expressed in the real world? We know that if leaves or potted plants are exposed to increasing CO_2, they indeed respond with increased growth. Many experiments have repeatedly demonstrated this effect. In fact, many of these studies support the idea that plant growth will increase by a factor similar to that estimated for photosynthesis, or 25%–40%. However, plants in nature are exposed to many factors that are difficult to replicate in a short-term experiment.

An experimental technique called Free Air CO_2 Enrichment (FACE) has been developed in which a plot of vegetation is surrounded by plumbing that emits CO_2 into the atmosphere. CO_2 sensors and anemometers monitor concentrations and winds to tailor the gas release to maintain the desired increase in concentration within the plot. More CO_2 is released on the upwind side than the downwind side, and the rate of release depends on the wind and mixing. This process sounds complicated, and it is, and also extremely costly, as CO_2 is quite expensive in the quantities required. This technology was originally employed for relatively short vegetation but has even been used in forests. FACE technology allows CO_2 enrichment of intact natural communities, but the technique has a number of challenges. For example, the concentrations increase abruptly rather than increasing over decades, as is occurring with global change. Because the process is expensive, the enrichment plots are small and few in number. Perhaps an even more important limitation is the total number of experiments. Compared with literally thousands of leaf and plant level studies, only a handful of FACE studies have been done, and none in old-growth, tropical, or boreal forests—likely sites for carbon storage.

One of the most complete studies was done in a southeastern U.S. loblolly pine forest by Duke University researchers. Multiple replicated plots were treated with high CO_2, and exhaustive measurements were made: how exhaustive is indicated by the fact that this one study has produced over 200 publications. The Duke study concluded that increasing CO_2 caused significant and sustained increases in plant growth and biomass, although it did not alter the relative distribution of biomass among roots, wood, and leaves. Within the experiment, there was considerable variability in soil fertility, with areas of higher and lower nitrogen availability. The impact of CO_2 varied within the experiment: growth responses were larger where nitrogen availability was higher, suggesting that in natural ecosystems the response to increasing atmospheric CO_2 will vary with other resources, such as soil fertility. Interestingly, over time, the growth responses were larger in wetter years than in dry years, in contrast with the hypothesis that the positive impact of CO_2 on WUE would help plants weather drought conditions.

After many decades of research on CO_2 fertilization, we can answer some questions. CO_2 does cause increases in growth and biomass accumulation, and seems to be more effective when nitrogen is plentiful. CO_2 does increase water-use efficiency, but its impact on plant growth under drought conditions is not clear. It is still not known whether CO_2 fertilization is responsible for global-scale carbon uptake, and if it is, whether this process will continue in the future. Despite some decades of creative thought, it is hard to come up with experimental approaches to unequivocally resolve this issue.

Increasing CO_2 in the oceans has a completely different type of direct effect on biology. Adding CO_2 to the atmosphere affects the chemical climate of the oceans, specifically its acidity, or pH. Oceanic CO_2 in surface waters is in rough equilibrium with the atmosphere, so that when its concentration increases in the atmosphere, there is a corresponding increase in the surface ocean. Although CO_2 in the atmosphere is chemically unreactive and is lost or converted to other forms only by the biospheres of land and the oceans, in the oceans CO_2 participates in complex reactions that govern the acidity and geochemistry of seawater. These reactions specifically influence the solubility of carbonate minerals needed by many marine organisms such as coral, shells, and skeletons to form structures.

In seawater, carbonate chemistry is governed by the following reactions:

$$CO_2(atmos) \rightleftharpoons CO_2(aq) + H_2O \rightleftharpoons H_2CO_3 \rightleftharpoons H^+ + HCO_3^-$$
$$\rightleftharpoons 2H^+ + CO_3^{2-}$$

Changes to atmospheric CO_2 equilibrate with the surface ocean within about a year. When CO_2 dissolves in seawater, it reacts with water to form carbonic acid (H_2CO_3), which dissociates to bicarbonate (HCO_3^-) and carbonate (CO_3^{2-}) ions. Increasing CO_2 increases the concentrations of all the carbonate species and also increases the hydrogen ion concentration and, conversely, decreases pH. As pH declines with increasing CO_2, then carbonate ion (CO_3^{2-}) concentrations decline, as the preceding equilibrium is driven to the left. This fact is important because the ocean's ability to absorb CO_2 depends on the extent of CO_3^{2-} dissolved in the water column, and because the ability of organisms to form carbonate structures depends on the saturation state of carbonate minerals. The carbonate ion affects $CaCO_3$ solubility according to the following equilibrium:

$$CaCO_3 \rightleftharpoons CO_3^{2-} + Ca^{2+}$$

The formation and dissolution of carbonate minerals are governed by the *omega* equation (Ω), where Ω denotes the saturation state of the seawater:

$$\Omega = \frac{[Ca^{2+}_{\cdot \cdot}][CO_3^{2-}]}{K'_{sp}}$$

where K'_{sp} is the apparent solubility constant of $CaCO_3$, which depends on temperature, salinity, pressure, and the particular solid mineral phase. Marine organisms utilize either calcite, aragonite, or high-magnesium calcite as primary structural material. The closer the seawater is to saturation, the easier it is for organisms to form structures. Many organisms even require supersaturation of $CaCO_3$ minerals to form shells. As seawater becomes increasingly acidic and undersaturated, it becomes more difficult for organisms to form structures.

As increasing CO_2 acidifies the oceans, the changing carbonate chemistry will have profound biological effects. Scientists realized these potential effects of acidity only in the late 1990s, and their magnitude is just now becoming apparent. Experimental studies (Fine and Tchernov 2007) showed that species of coral grown in acidified waters completely lost their carbonate skeletons but regrew them when returned to normal waters. Many studies are now documenting the impact of decreasing saturation state (Ω) on coral growth. On the Great Barrier Reef of Australia, calcification rates of corals declined by more than 20% in a 5-year period, although some of this effect was due to other pressures, such as pollution and warmer temperatures.

Many groups of plankton form calcareous structures and will be affected by acidification. These groups include the coccolithoforids, foraminifera, and some pteropods. These groups form shells or other structures from (mainly) aragonite or calcite. Experimental studies of planktonic organisms show a variety of responses to pH: decreases in shell mass, decreased overall size, malformed coccoliths (shells of coccolithoforids), and reduced growth rates (Doney et al. 2009). Within this overall picture, there is considerable species-level (and even strain-level) variability. With increasing CO_2 some species show growth increases, and some data from marine cores documenting the history of plankton response to CO_2 suggest a fair amount of adaptive plasticity, at least up to present-day levels. Ecologically, it is unclear how changes to calcification, and differences between species in sensitivity to acidification, might affect competition between groups and thus affect the organisms that feed on them. It is likely that acidification will have significant consequences for marine ecosystems, but at this point, we can't predict what they will be.

Based on the evidence that increasing CO_2 and ensuing changes to pH and Ω could cause major changes in marine ecosystems, what can the past tell us? In the extreme, acidification could cause massive die-off and extinctions of some groups of marine organisms. Fifty-five million years ago a large rapid increase in atmospheric CO_2 occurred, the Paleocene-Eocene Thermal Maximum (PETM), which led to changes in ocean pH and carbonate chemistry (Doney and Schimel 2007). Although the PETM had major climate and ecological consequences, the only taxa known to have gone extinct were within the benthic foraminifera, and this extinction may have resulted from changes to oxygen status rather than pH. However, there were many other ecological and geochemical differences between the world of the PETM and the present, and so the analogy is not perfect.

Ocean acidification has complicated another area of climate change. In recent years, many people have questioned what could be done if greenhouse gas–caused warming temperatures and changing climate begin to have serious consequences, but reduction of energy-related emissions remains difficult. There are several proposals for ways to *engineer* the climate (called *geoengineering*; Russell et al. 2012) to cool temperatures without reducing CO_2 and other greenhouse gases. These include adding artificial aerosols to the atmosphere to reflect sunlight and cool Earth's surface. However, managing the climate by reducing incoming solar reduction (*solar radiation management*) while allowing atmospheric CO_2 concentrations to increase would lead to increasingly serious ocean acidification. These connections among the carbon cycle, the oceans, climate, and geoengineering are just one example of the importance of taking an Earth Systems view of the planet rather than considering processes and *spheres* (*biosphere, hydrosphere, etc.*) separately.

LAND USE CHANGE AND DEFORESTATION

Now, let's revisit global analyses. Recall that atmospheric data suggest midlatitude Northern Hemisphere uptake of carbon, whereas CO_2 fertilization theory predicts globally distributed uptake. The Duke result even suggests that the tropics, where nutrients are often not limiting, should have high uptake, yet the atmosphere indicates this is not occurring. This paradox has led many researchers to look for other mechanisms that might cause carbon uptake in the Northern Hemisphere.

The obvious mechanisms relate to land use and management. Several ideas suggest themselves. First, vast areas of forest were harvested in the Northern Hemisphere during the Industrial Revolution and after, but many of those forests are now managed very differently. These forests, cut in the 1800s and 1900s, could still be gaining biomass and storing carbon. In addition, in some areas of the Northern Hemisphere, land cleared for agriculture has been abandoned owing to loss of soil fertility, changing economics, and changing cultural values. These areas frequently revert to forest cover and may be gaining carbon. We have discussed nitrogen limitation of growth, yet much of the Northern Hemisphere receives significant additional nitrogen from air pollution, which could be increasing growth.

All these mechanisms could be causing carbon uptake now, for reasons unrelated to climate or CO_2 feedbacks. Many researchers have carefully analyzed forest data, collected mostly for forest management purposes, and have carefully estimated carbon balance from these data. These studies increasingly show Northern Hemisphere forests to be gaining carbon, mainly for reasons related to forest dynamics, such as maturation of young trees, effects of increased nitrogen availability, reduced losses from wildfire control, and growth of trees on former farmland. These mechanisms could be responsible for the bulk of the carbon uptake observed today, with increasing CO_2 making a secondary contribution. If today's uptake is due to forest ecosystem processes (and not mostly a CO_2 effect), it could still vary greatly from year to year with climate around an average level set by these forest processes, as we observe in the atmosphere. However, we can't predict the magnitude of future uptake simply from climate and CO_2 values but, instead, will have to understand global forest ecological dynamics with some precision.

If forests are growing because of historical land use, then this might not continue into the future, or might require careful land management to be sustainable. Some refer to the potential that the 1–2 Gt C that terrestrial ecosystems now store each year will decline as "sink saturation." Sinks refer to mechanisms that remove CO_2 from the atmosphere and saturation suggests that these mechanisms will saturate (as the CO_2 fertilization effect does). If that were to occur, then the growth rate of CO_2 could increase dramatically, with a concomitant increase in the rate and severity of climate change. Is there a *tipping point* in the terrestrial carbon cycle where uptake would decrease or even change to release? Some ecologists, notably George Woodwell, think this possible, and some global carbon cycle models show behavior consistent with this fear.

The flip side of forest regrowth is deforestation. Deforestation—the cutting or harvesting of forests for fiber or fuel and to clear land for agriculture—releases carbon because ecosystems with high carbon storage (mainly in wood) are replaced by systems with low carbon storage. This situation may be permanent, when forests are replaced by arable land, or temporary, when forests are harvested but regrow naturally or as managed plantations. How much carbon is released depends on how the wood and other forest products are used. If land is cleared for agriculture, most of the wood may be burned and returned to the atmosphere within a few years. If forests are harvested for timber, then the wood produced may last for years or decades and will only slowly be returned to the atmosphere. Soil carbon storage may also change. Keeping track of the fate of forests requires complex bookkeeping: How much land area is converted? How much carbon was stored initially and how much remains? Is the land used for agriculture or does it return to forest? Is the timber used for wood products, burned, or otherwise disposed of? Because many forest-clearing activities take place on very small parcels of land, keeping track of deforestation impacts on the carbon cycle requires vast amounts of information, an approach pioneered by Richard (Skee) Houghton in the 1980s (Houghton 1983).

Today's estimates of deforestation draw on remote sensing of vegetation, national inventories of forest resources, surveys of biomass, and increasingly sophisticated instrumental and space-borne measures of biomass (Saatchi et al. 2011). Since the beginning of the industrial period about half of the world's forests have been deforested, accounting for nearly 20% of human-caused emissions to the atmosphere. Deforestation rates may have peaked globally in the 1990s at about 1.5 (± 0.7) Gt C/yr. In the latter part of the first decade of this century, the rates appear to be closer to 0.9 (± 0.7) Gt C/yr. The high uncertainty reflects the difficulties of estimation, the vast amount of land area involved, the diversity of initial carbon levels, and postdeforestation practices.

Deforestation has changed over the decades. High rates of forest clearing began in Europe and North America, and spread to temperate Eurasia. Tropical deforestation began after the mid-twentieth century, but rates of tropical deforestation did not extend beyond the midlatitudes until the 1970s. Deforestation became widespread in Southeast Asia, Africa, and South America in response to growing human populations and changing economic incentives. Early tropical deforestation was driven by agricultural practices that were sustainable at low human population levels but became nonsustainable as populations began to grow. As commerce became globalized, the tropics began to export agricultural commodities, and today, a high proportion of deforestation supports global markets in wood, forest products, and agricultural commodities grown on land converted from primary forest. The post-1990s decreases in deforestation rates reflect growing pressure to protect primary forest, protect biodiversity and carbon stores, and preserve indigenous ways of life. Tropical forests remain at risk, and the current rates of deforestation, although perhaps a

125

third to half of peak rates, continue to endanger carbon stocks, biodiversity, and indigenous cultures.

Deforestation remains a significant contributor to climate change (via the carbon cycle), but it affects the Earth System differently than does fossil fuel burning. Recall that burning fossil fuel releases carbon fixed millions of years ago, carbon that would have remained stored for millions of years more, and so increases the amount of carbon in the atmosphere–ocean–land components of the Earth System. Whereas burning fossil fuel creates a change in the carbon cycle that requires many thousands of years to return to steady-state conditions, deforestation basically short-circuits the normal processes operating within the land–atmosphere–ocean system. If deforestation were the only source of CO_2 in the atmosphere, the system could return to equilibrium much more quickly. Reducing deforestation, and reforesting in the tropics, can contribute significantly to reducing climate impacts of the altered carbon cycle, assuming that tropical climate remains stable enough to support the vast preindustrial areas and biomass levels of tropical forest.

THE SURFACE ENERGY BALANCE AND CLIMATE

The effects of deforestation and other changes to land cover extend beyond the carbon cycle. For many years, meteorologists thought that land ecosystems were passive participants in climate, absorbing moisture from precipitation and energy from the sun, and releasing it back to the atmosphere in response to its state (temperature, humidity, and so on). In early mathematical models of climate, the land was defined by only two parameters, albedo and water-holding capacity, or how many centimeters of moisture the soil could hold, for evaporation back to the atmosphere. Many people have an intuitive (and correct) contrasting view, that vegetation affects climate, and that changing vegetation alters the climate. This opinion probably results from our individual perception of climate as temperature and humidity. We know, for example, that when we enter a forest, the temperature drops, and it may feel more humid. Pioneers in North America believed that "rainfall followed the plow" and that the conversion of the Great Plains prairies to cropland would increase rainfall.

Early climate modelers knew about forest microclimates and other local effects of vegetation but believed that any active effects of biology would be overwhelmed by large-scale factors in the atmosphere. Three-fourths of Earth's surface is covered by oceans, and at the dawn of climate modeling, scientists viewed the Earth as fundamentally an ocean planet, and understanding atmosphere–ocean problems as the main challenge.

An understanding of the effects of vegetation and changes to vegetation on climate originated from climate models rather than from observations. The large scales over which the atmosphere and ecosystems interact, and the primitive measurement technology available decades ago, made it very hard to study vegetation–climate interactions observationally or experimentally, so modelers began by incorporating biological first principles into simple models of the land biosphere. The sense in mainstream climatology that the land biosphere couldn't participate actively in climate was so strong that only a few researchers pursued the idea.

Early land surface models simulated the two parameters mentioned previously, albedo and soil moisture. Modelers addressed albedo by assigning different values to different vegetation types, based on observations, which allowed changing vegetation or mosaics of vegetation at the land surface to create diversity in energy balance in time and space. Early models took a simple approach to representing ecosystem control over water balance. Modelers imagined the land surface as a bucket containing water whose depth equaled the amount of water the soil could hold. They imagined the landscape covered by buckets containing water of varying depths (although some very early models covered Earth with buckets of the assumed global average soil moisture). If precipitation occurred in excess of the amount the bucket could hold, it was assumed to be runoff. The runoff was initially assumed to be irrelevant to climate and either vanished or returned directly to the ocean. The water in the buckets could then evaporate in response to the atmospheric climate. Variability in buckets in space could create patterns in atmospheric water vapor and latent heating.

Several early studies used this type of model to examine the possible effects of variability in the land surface on the overlying atmosphere. Adjacent surfaces of contrasting temperatures tend to create a circulation. Air rises over the warmer surface, and cooler air from the adjacent cooler surface flows in to replace the rising air. This process is most familiar along coastlines. At night, the land cools faster than the ocean because of the ocean's higher heat storage capacity, so winds blow offshore, from land to sea. During the day, the opposite effect occurs, and onshore breezes blow, from sea to land. Could contrasts between vegetation types produce similar circulations? Model calculations by researchers such as Roger Pielke Sr. and Rick Anthes in the 1970s and early 1980s showed that oases in the desert, and contrasts between grasslands and irrigated crops, produced significant differences in the land surface temperature because of differences in evaporation rates. The differences, due to the cooling effect of evaporation, were predicted to be large enough to cause sea breeze–type circulations.

These sea breezes arise largely because of modeled evapotranspiration as a control over land surface temperature. Sea breezes caused by land surface temperature contrasts are typically fairly weak, and may not even be evident when regional winds are strong. Accordingly, developing observational support for the sea breeze hypothesis was difficult, although surface and airborne observations provided early confirmation of the differences in temperatures between different vegetation types.

Whereas early studies focused on small-scale effects of land surface variability, the prevalence of large-scale human modification of the landscape was becoming increasingly evident to scientists. If local changes such as the creation of irrigation districts could affect local weather, what effects would wholesale changes to vegetation have? As land surface models became more sophisticated biologically, and observations of water and energy fluxes above more and more types of vegetation became available, scientists realized that different vegetation types, even in a common climate, could affect the way energy was partitioned between sensible and latent, and could also control the seasonal timing of latent heat exchange.

Conversion of grasslands to croplands affected the storage of water from winter snow and rain, and changed the timing of maximum growth. Pielke showed that such conversion led to conditions more conducive to severe storms along so-called atmospheric drylines (a boundary between more and less humid air over the continent; Ziegler, Lee, and Pielke 1997). The exact mechanism of land use effects on severe weather are complex, as increases in atmospheric moisture increase convective instability, promoting stormy conditions, but reductions in sensible heating can reduce the formation of deep cumulus clouds and thunderstorms. Increased evapotranspiration may increase rainfall when the atmosphere is dry but may actually decrease rainfall when the atmosphere is humid, by reducing sensible heating needed to trigger convection. Both landscape patterns and changes to these patterns by land use have significant effects on weather and climate, and the more these influences have been studied, the more significant they seem to be.

The water and energy balances of the Amazon basin have received a great deal of attention from climatologists. Early observational data suggested that the biosphere and the atmosphere are tightly coupled there, and the high rates of deforestation there are well known, suggesting that land use change in the Amazon could affect regional or possibly even global climate. The Brazilian hydrologist Eneas Salati (Salati and Vose 1984) first revealed a strong role for the biosphere in the Amazon Basin's water cycle, based on isotopic evidence. The oxygen in water contains both the more abundant isotope ^{16}O and the rare heavy isotope ^{18}O. As water moves over the continent, it equilibrates between the vapor and liquid phases. When water evaporates (changes from liquid to vapor), the light isotope evaporates faster than the heavy isotope, leaving liquid water enriched in ^{18}O. On most continents, the farther inland from the oceanic source of water, the more depleted in the heavy isotope ^{18}O precipitation becomes.

Salati showed that rainfall in the Amazon showed only very slight change in ^{18}O compared with that on most continents. One mechanism for this occurrence would be for much of the water precipitated from the atmosphere to return to the atmosphere more or less where the rainfall occurred. If this was true, then the isotopic imbalance created by the precipitation would be rebalanced by return of those waters (assuming that evaporation at the land surface did not additionally change the isotope ratios). Salati made extensive measurements of the isotopic composition of water in the Amazon basin that led to the *recycling* hypothesis.

In this hypothesis, water falling on the Amazon basin is quickly returned to the atmosphere by evapotranspiration, reflecting the abundant forest, high leaf area, and long growing season. In fact, in their 1984 publication, Salati and Vose showed that in the Amazon, the residence time of water in the plant–soil system was only 5.5 days. They imagined the atmosphere above the Amazon as a kind of conveyer belt. Water from the ocean enters the Amazon region, and some of it falls as rain. That air parcel then moves inland and is resupplied with water from evapotranspiration from the forest it passes over. Precipitation again occurs, with about half of the rainwater coming from the atmosphere and about half from the previous area of rainforest. As air passes over the Amazon basin, any given water molecule may be recycled seven times before the air once again leaves the continent.

Because of this tight coupling, the atmosphere and biosphere are tightly linked over the Amazon basin. Most of the water returned to the atmosphere is associated with either plant water use (transpiration) or evaporation from water intercepted by the plant canopy and evaporated directly from the leaf surfaces. If the Amazon rainforest were to be replaced by crops or pasture grasses, then this tight coupling could be disrupted. Less forest cover could imply less evapotranspiration and, hence, less rainfall. This could lead to disruption of the ecosystem, which depends on high rainfall rates to maintain productivity and nutrient cycling, and could create a positive feedback, leading to a drier Amazon.

Salati's early work using isotopes, and the growing awareness in the 1980s of Amazonian deforestation, led many researchers to try to quantify the effect of Amazon vegetation on climate using climate models. Some of the earliest modeling experiments on atmosphere–ecosystem coupling were motivated by this issue. Since the atmosphere and ecosystems interact over vast distances to control rainfall, small-scale deforestation experiments or even the ongoing rainforest destruction by settlers could not create the conditions to test this hypothesis. However, paired simulations of the climate system can be set up, with one experiment simulating the Amazon as covered by rainforest, and the second, vegetation representing herbaceous crops or pastures, holding everything else constant. These experiments can capture effects of the different rooting depths of trees versus herbaceous vegetation, effects of leaf area (higher in the forest), seasonality (forests are perennial, while crops are harvested), and other functional differences. Then, the modeled temperatures and rainfall rates can be compared between the two experiments. Such model experiments are critical for studying Earth System processes that occur over scales too large to simulate with actual manipulative field experiments.

In one such model experiment (Shukla, Nobre, and Sellers 1990), surface temperature increased by 2.5 °C, and precipitation decreased by 26%. Not only did rainfall decrease, but the length of the dry season increased. Many plants are able to survive seasonal drought, but a systematic increase in the length of seasonal dry period could kill some plants, prevent others from reestablishing, and disrupt the complex and coupled life cycles of plants, pollinators, and other animals that maintain the rainforest. Shukla and his colleagues concluded that "changes in the region's hydrological cycle and the disruption of complex plant–animal relations could be so profound that, once the tropical forests were destroyed, they might not be able to reestablish themselves." Subsequent studies have supported early work on the effect of forests on the Amazon's climate and, as we'll see, added the dimension of the carbon cycle to the hydrological cycle impacts.

The Amazon example showed that removing vegetation would lead to a warmer surface (because of less evaporation). Subsequent Amazon experiments caused even more warming because they also included the effects of the CO_2 lost to the atmosphere from deforesting the basin. Could changing vegetation ever lead to warmer conditions? Gordon Bonan (2008) did an early experiment, similar to the Amazon one, in which he compared simulations of climate including and removing the boreal forest. Again, this is an effect that occurs over areas too large to test with actual experiments, so models play a crucial role.

In Bonan's experiment, the effect of removing the boreal forest was the opposite of that produced by removing the Amazon rainforest. In the boreal, the main effects occurred during the winter. During the Northern winter, snow covers the ground, but the prevailing coniferous evergreen trees retain their needles. The trees have low albedo and are relatively dark in color and absorb sunlight, warming the surface. When the trees are removed, the bright snow reflects most of the incoming sunlight and cools the surface. As a result, removal of trees in high-latitude forests might actually cool the climate a bit by replacing the dark forest with bright snow.

Eyal Rotenberg and Dan Yakir (2010) showed that carbon cycle and energy balance effects can have opposing effects in dryland forests. The Yatir Forest is a large restored forest in Israel that replaced desert shrubland. Rotenberg and Yakir observed water, carbon, and energy fluxes there for many years, and also examined satellite maps of surface temperature. As it grew, the Yatir Forest took up carbon, and at surprisingly high rates given its dry environment. The carbon impact of forest growth tends to cool the global environment, by removing atmospheric CO_2. Rotenberg and Yakir estimate that this produces an effect similar to that seen in the boreal. The desert is bright and has high albedo, reflecting sunlight. The forest has an albedo about half that of the desert it replaces and so absorbs more of the incoming solar radiation, resulting in surface heating. While carbon uptake by the growing Yatir Forest is a negative feedback to climate, the change in albedo is a positive effect. The tradeoff time (Rotenberg and Yakir 2010) is about 30–50 years, based on how much carbon the forest must absorb to balance the albedo effect. The Yatir study was based on extrapolation from measurements in an afforested region on scales large enough to affect the atmosphere but represents what could happen in the extensive dryland regions of the world.

METHANE AND NITROUS OXIDE

Carbon dioxide is only one of several important trace gases affecting global climate. Although CO_2 is the most abundant and important in radiative forcing in the atmosphere, methane (CH_4) and nitrous oxide (N_2O) are also of great importance at 1800 and 325 ppb respectively, as of 2011. Methane and nitrous oxide are increasing in the atmosphere today and also generally paralleled CO_2 in the ice core records of glacial–interglacial cycles. The current contribution of CH_4 to radiative forcing is about a third that of CO_2. Nitrous oxide is present at a lower concentration but is a more effective absorber and so, despite its low concentration, still accounts for about 10% as much forcing as CO_2.

Despite the very different biogeochemical pathways controlling atmospheric CO_2, CH_4, and N_2O, their correlations suggest they respond to environmental change in related ways. Methane and nitrous oxide are very effective greenhouse gases, and both are, molecule for molecule, more powerful in trapping heat in the atmosphere than is CO_2. Methane, for example, is 72 times as effective as CO_2, and N_2O is about 310 times stronger than CO_2. While CH_4 and N_2O differ from CO_2 in many ways, one of the most important is that both have well-defined atmospheric lifetimes and are (largely) destroyed by chemical oxidation in the atmosphere, although both have other minor sinks. The lifetimes of N_2O and CH_4 are described by the exponential law described earlier. Methane has about a 12-year lifetime and does not recycle within the Earth System as CO_2 does but, rather, is oxidized in the atmosphere to CO_2. Nitrous oxide has a 114-year atmospheric lifetime, and it is also oxidized in the atmosphere. While CO_2 is chemically inert in the atmosphere, CH_4 and N_2O are reactive compounds. As a result, their feedbacks in the Earth System operate via atmospheric chemical interactions.

METHANE

Methane has closely tracked CO_2 in the ice cores, so that you might think their cycles were tightly coupled. However the sources and sinks of CH_4 are very different from from those of CO_2. Whereas the glacial–interglacial cycles of CO_2 are largely due to interactions with the oceans, CH_4 is emitted to the atmosphere mainly on land (only 1% –2% of emissions come from the oceans). Methane is produced from biologically fixed carbon when it is decomposed by microorganisms of the group Archaea, under low-oxygen conditions. Methane production, *methanogensis*, is a form of anaerobic respiration (respiration in the absence of oxygen). In anaerobic respiration an alternative terminal electron acceptor must replace oxygen. In methanogenesis, either CO_2 or a simple organic carbon compound serves as that terminal electron acceptor, and CH_4 is then a product of the reaction:

$$CO_2 + 4H_2 \rightarrow CH_4 + 2H_2O \text{ (CO}_2 \text{ as terminal electron}$$
acceptor)

$CH_3COOH \rightarrow CH_4 + CO_2$ (organic matter as terminal electron acceptor)

Methanogenesis occurs in environments that are low in oxygen but have available carbon as a substrate for the methanogenic organisms. Globally, methane is produced in a variety of anaerobic habitats. Tropical and high-latitude wetlands are important, as are rice paddies (which are, in effect, agricultural wetland systems). Methane is produced in the guts of ruminants (cows, sheep, goats), where their feed is digested anaerobically. Other creatures, such as termites, also produce CH_4 anaerobically in their gut. Methane is also released from biomass burning, and so CH_4 accompanies the various other combustion products during fires. Methane is also produced in marine sediments, although relatively little of the methane produced in anaerobic marine habitats is emitted to the atmosphere; it is mainly oxidized in sediments and the water column before it reaches the ocean surface (Reeburgh 2007). Aside from its ecosystem sources, some methane is also released during energy production, vented as natural gas from wells or in leaks from natural gas wells and transmission pipelines. Methane is also produced by anaerobic metabolism in the *disgustosphere* and is emitted from landfills, waste treatment plants, and from untreated sewage. Thus, a scientist interested in global methane has to have a very broad range of expertise!

Although CH_4 from all these sources affects the climate, some of them may participate in climate feedbacks in which changes to climate will trigger changes to emission rates. Tropical wetlands produce a significant amount of methane, perhaps a sixth of the global total, with another 8% or so coming from high latitudes.

Why are wetlands so hospitable an environment for methane production? Remember that methanogenesis requires anaerobic conditions. Water is an effective barrier to oxygen, as O_2 diffuses across the air–water interface very slowly. Even a small amount of biological activity will consume most of the available O_2 initially available in water, and because diffusion is slow, the water column in wetlands will typically be anoxic (without oxygen) to within a few centimeters of the surface. As decomposition of organic matter proceeds, substrate for methanogens is produced, including acetic acid and CO_2, typical substrates for methanogens. In productive tropical wetlands, where plant growth continues for much of the year, there is a constant supply of organic matter, and methane may be produced in large amounts. When methane reaches high concentrations, it is energetically favorable as a substrate for oxidation by alternative electron acceptors such as sulfate (SO_4^{2-}) ions. Anaerobic oxidation of methane may then take place and can consume significant amounts of the gas before it is emitted to the atmosphere.

Methane metabolism does not provide a great deal of energy, and after oxygen is depleted, other electron acceptors such as NO_3^- and SO_4^{2-} ions are used preferentially. Methanogenesis may not begin until those substrates are also depleted. In marine sediments, where surface pore waters have high SO_4^{2-} concentrations, methane is produced even deeper in the sediments. Since gases diffuse slowly through water, if methane must diffuse through the water column in the wetland, much of the gas may be consumed as it passes through the water. Some of the methane may transit the water quickly if bubbles form and rise rapidly to the surface. Another important pathway is via plants growing in the wetland, many of which have special tissue, called *aerenchyma*, that allows oxygen to be transported to their roots. Numerous compounds produced under anaerobic conditions are toxic to plants and also inhibit nitrogen availability. Oxygen diffuses from the roots of aerenchyma-containing plants and creates microzones of oxic conditions that aid their metabolism. However, these air plants also allow methane to diffuse rapidly through the gas-filled aerenchyma spaces, which provide a shunt, allowing methane to pass directly to the atmosphere without passing through a zone of oxidation. This increases emissions to the atmosphere in vegetated wetlands.

Changes to the temperature (affecting rates of microbial metabolism) and water (affecting the duration of flooding and anaerobic conditions) could clearly affect the emissions of methane from these ecosystems. In many wetland ecosystems, primary productivity is also correlated with methane production owing to both organic matter inputs and fluxes through transpiring plants, which suggests that changes to tropical or boreal NPP could affect methane production. In high latitudes, many wetlands lie atop permafrost layers, which are often rich in organic matter. If these permafrost soils thaw, then the vast carbon stocks they contain would once again be available for microbial decomposition and could affect methane and CO_2 emissions from these systems.

The atmospheric increase in methane results from an imbalance between sources of methane (emissions from wetlands, animals, and other sources) and destruction of atmospheric methane via oxidation to CO_2. There are two major pathways for methane oxidation: atmospheric reactions (the largest) and oxidation in soils. In the atmosphere, the primary oxidation reaction of methane begins with a collision with $OH\cdot$, the hydroxyl radical:

$$CH_4 + OH\cdot \rightarrow CH_3 + H_2O$$

Following this reaction, a complex chain of reactions ensues in which formaldehyde (CH_2O) is typically formed. Formaldehyde is oxidized to carbon monoxide, and then, finally, CO is oxidized to CO_2. Methane oxidation is one of the most important chemical reactions in the atmosphere, and because of the abundance of CH_4 relative to most other atmospheric organic compounds, it competes with other organic compounds for oxidizing agents and so influences the atmospheric lifetime of many other gases, including itself. As methane becomes more abundant, its oxidation consumes an increasingly larger fraction of tropospheric $OH\cdot$. As the availability of $OH\cdot$ radicals decreases in the atmosphere, their rate of attack on CH_4 decreases, which increases the atmospheric lifetime of CH_4 and so increases its effect on climate in a coupled physical-chemical-biological feedback loop. Most CH_4 is oxidized in the lower atmosphere, but some is oxidized in the stratosphere. The stratosphere is extremely dry, and the water released in the initial oxidation step of methane is an important source of water in the stratosphere, affecting its temperature. Climate changes that affect CH_4 production can have additional feedback effects on the climate by affecting stratospheric water.

The atmospheric chemical sink dominates the destruction of methane, but more recently, scientists have discovered a second sink in soils. Soil consumption of CH_4 was first reported by Harriss, Sebacher, and Day (1982), who observed concentration decreases during some seasons in flux chamber headspaces in a swamp. Keller, Mitre, and Stallard (1990) reported similar decreases in moist tropical forest soils. Methane oxidation activity in soils is reduced or eliminated by disturbance or nitrogen fertilizer additions (Mosier et al. 1991; Steudler et al. 1989; Reay and Nedwell 2001). The impact of nitrogen appears to vary among different soils, perhaps owing to different microbial populations contained in them (Gulledge, Doyle, and Schimel 1997). Enzymes involved in NH_4^+ oxidation to NO_3^- are similar to those involved in CH_4 oxidation, so it has been hypothesized that increased nitrogen deposition (from air pollution) and agricultural fertilization of soils could reduce the soil methane sink. Globally, agricultural development may not only increase CH_4 emissions but may also decrease CH_4 oxidation and so increase the lifetime of CH_4 as a climate agent.

Nitrous oxide (N_2O) is the third critical biogenic trace gas. Nitrous oxide is produced by two key nitrogen cycle processes. Nitrification is the oxidation of ammonium (NH_4^+) to nitrate (NO_3^-): this pathway produces N_2O as a by-product and is carried out by two dominant groups of bacteria:

$NH_3 + CO_2 + 0.5O_2 \rightarrow NO_2^- + H_2O + H^+$ (mediated by *Nitrosomonas*)

$NO_2^- + CO_2 + 0.5O_2 \rightarrow NO_3^-$ (mediated by *Nitrobacter*)

As you can see, nitrite (NO_2^-) is an intermediate in this set of coupled microbial transformations. A small fraction of the NH_3 converted to NO_2^- is lost as N_2O as a by-product of the complex enzyme-bound steps between NH_3 and NO_2^-. Nitrification typically occurs in well-aerated soils, with adequate moisture to allow for vigorous microbial activity but not so much water that oxygen diffusion becomes limited; it also occurs in some zones of the ocean. NH_4^+ availability also limits nitrification, since an ammonium ion is the substrate for the reaction. As a result, N_2O production tends to be correlated with overall nitrogen cycling rate, and more N_2O will be released from systems with high nitrogen availability. As a result, systems such as fertilized agriculture and tropical rainforests have high nitrification and N_2O emission rates.

Denitrification reduces nitrate ultimately to dinitrogen gas (N_2). As in nitrification, N_2O is produced as an intermediate, but at much higher rates than in nitrification. The reaction for denitrification, showing N_2O as an intermediate, is

$$NO_2^- \rightarrow NO + N_2O \rightarrow N_2$$

The reaction for denitrification when it goes to completion to produce N_2 is

$$2NO_3^- + 10e^- + 12H^+ \rightarrow N_2 + 6H_2O$$

As can be seen from the requirement for electrons, denitrification requires reducing conditions. In soils, denitrification occurs under wet or waterlogged conditions. While denitrification to N_2 is most rapid under waterlogged conditions, the loss of nitrogen as N_2O tends to occur under slightly drier conditions. Recall that gas diffusion is slow in water, and so in fully waterlogged soils, N_2O is also slow to diffuse out of soils. When soils are at intermediate water levels, they contain a mix of soil pores full of water, which are an ideal environment for denitrification, and air-filled pores that allow N_2O to diffuse out before being fully reduced to N_2. Because the physical structure of soils is important, different soil types often have different denitrification rates. Fine-textured soils (clay and silt soils) have higher rates than sandier soils with larger soil particle sizes and consequently larger pores. Like nitrification, denitrification also is most rapid when nitrogen availability is high. Thus, it is also most important in soils with high nitrogen availability and cycling rates.

Like methane, nitrous oxide is mainly destroyed by photochemical reactions in the atmosphere. It is inert in the troposphere and so is oxidized only when it reaches the stratosphere. Stratosphere–troposphere exchange is relatively slow, and N_2O destruction is also relatively slow, so N_2O has a long atmospheric lifetime, contributing to its large effect on climate.

In the stratosphere, ozone is produced from O_2 by reaction with sunlight ($h\nu$) according to the Chapman (1930) reactions:

$$O_2 + h\nu \rightarrow O + O$$

$$O + O_2 \rightarrow O_3$$

Ozone in the stratosphere is critical to protecting Earth from excessive UV radiation. Some of the N_2O is oxidized by the oxygen produced in the ozone reaction discovered by Paul Crutzen in 1970 (and for which he won the Nobel Prize in Chemistry):

$$N_2O + O \rightarrow 2NO$$

The nitric oxide (NO) produced then interacts with ozone:

$$NO + O_3 \rightarrow NO_2 + O_2$$

$$O_3 \rightarrow O + O_2$$

$$NO_2 + O \rightarrow NO + O_2$$

This is an extraordinarily important reaction, because the net reaction for this process is

$$2O_3 \rightarrow 3O_2$$

That is, the net effect of the oxidation of N_2O is to destroy stratospheric ozone. Although we think of chlorofluorocarbons as the main destroyers of ozone, the human-caused increase in N_2O is also to blame. Nitrous oxide is the largest natural sink for ozone, so that in preindustrial conditions the lifetime of stratospheric ozone was determined mainly by N_2O. Ozone warms the stratosphere, by absorbing incoming solar ultraviolet radiation and upwelling infrared radiation from the troposphere. Decreased stratosphere ozone results in cooler temperatures. Over recent decades, the mid- to upper stratosphere has cooled by 1 °C to 6 °C at the same time that greenhouse gas amounts in the troposphere have increased, probably owing to reductions in stratospheric ozone.

Both CH_4 and N_2O have major impacts on the Earth System as greenhouse gases, through primary chemical effects and through secondary climate effects. In the case of CH_4, aside from its direct effect on climate as its concentration increases, not only is its own lifetime in the atmosphere prolonged but so are the lifetimes and hence climate and pollutant effects of all other compounds attacked by OH·. Because OH· is the major reactant that destroys pollutants in the atmosphere, this is a major effect. In addition, the oxidation of CH_4 in the troposphere leads to increased tropospheric O_3 under many conditions, and tropospheric O_3 acts as an additional greenhouse gas, increasing the effects of increasing CH_4. Thus, if changing climate causes CH_4 to change, the effects will be multiplied throughout the Earth System, changing the chemical environment, the lifetime of atmospheric pollutants, and causing additional local climate change through the ozone cycle. In addition, by adding water to the stratosphere, CH_4 affects the upper atmospheric climate and is responsible for further Earth System consequences. Nitrous oxide is also interwoven through many Earth System processes and contributes to another global change, stratospheric ozone depletion. As climate and land management practices cause N_2O to continually increase in the atmosphere, its climatic and chemical effects will continue to increase. Ecosystems are intimately tied into the chemistry and climate of the planet, and increasingly, the work of ecologists, atmospheric chemists, and climatologists addresses parts of a seamless and highly interconnected planetary system.

DUST, AEROSOLS, AND DIMETHYLSULFIDE

Atmospheric particulates (aerosols) play a major role in the climate system. Aerosols typically cool the planet by reflecting sunlight back to space. This is one way that volcanic eruptions affect climate, as discussed earlier. While volcanic and industrial aerosols affect climate and hence ecosystems, particulates from biological sources also result in aerosol-mediated feedbacks between ecosystems and climate.

One particularly important potential feedback is described by the CLAW hypothesis, named after the initials of its proposers, Charlson, Lovelock, Andreae, and Warren (Charlson et al. 1987). The hypothesis proposes a feedback loop between available energy, either as heat or increased sunlight, and the growth rates of phytoplankton

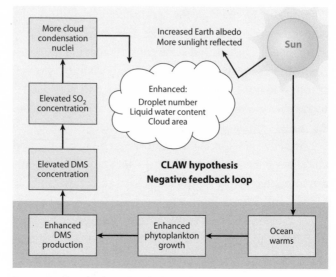

Figure 14. The Charlson-Lovelock-Andreae-Warren (CLAW) hypothesis for the modulation of climate energy balance based on a feedback loop between oceanic biogeochemistry, trace gas fluxes, and aerosols.

(Source: Charlson et al. 1987)

(see figure 14). Certain groups of phytoplankton, such as coccolithophorids (and other marine organisms), synthesize dimethylsulfoniopropionate (DMSP), and as their growth rates increase, so does the production of DMSP. DMSP is microbially transformed to dimethyl-sulfide (DMS) in seawater and subsequently diffuses out to the atmosphere. In the atmosphere, DMS is oxidized to sulfur dioxide, which acts to condense aerosol droplets. These aerosols act as cloud condensation nuclei (CCN) and increase the cloud droplet number, which in turn increases the liquid water content of clouds and the cloud area. The resultant increased cloud albedo reflects additional sunlight back to space, which tends to cool the surface. The reduction in available energy, either as sunlight or heat, closes the feedback loop.

This feedback could stabilize climate. Warmer or sunnier (less cloudy) conditions boost DMSP production through increased phytoplankton growth and as a result of phytoplankton stress responses to UV radiation, increasing DMS in the marine boundary layer (MBL). DMS is oxidized to SO_4^{2-}, which increases, through a series of chemical and physical steps, cloud albedo. DMS is a major source of SO_4^{2-} in the MBL, and in the present climate, significant fluxes of DMS are affecting the concentration of SO_4^{2-}, which is a requirement for operation of the feedback loop. Today after 24 years, the CLAW hypothesis remains unproven but has inspired an extraordinary amount of research. As of a 1997 review paper, 700 papers had addressed different aspects of the hypothesis, and as of a 2008 review paper, this number had grown to 1500 papers (Ayers and Cainey 2008). Despite all this research, as Ayers and Cainey note, "the coupled systems inherent in the CLAW hypothesis are so complex that we have yet to learn 'enough to give a defensible answer'".

Some of the complexity arises from the physiology of DMSP in phytoplankton. DMSP breakdown to DMS may reduce stress on organisms due to UV radiation, increasing surface temperatures, or nutrient limitation, although definitive evidence for an antioxidant role remains elusive. DMSP may also play a role in plankton regulation of osmotic potential and in sulfur metabolism. The underlying mechanism of the CLAW hypothesis, as framed by Ayers and Cainey, is as follows:

> Put even more simply, the hypothesis is that oceanic phytoplankton when pushed to either extreme of their coping range (too hot and sunny or too cold and dim) will respond by altering their dimethylsulfide (DMS) emissions so as to decrease or increase solar input to the ocean surface by regulating marine stratiform cloud reflectivity, hence driving the system back towards the middle of the phytoplankton coping range.

Lacking clarity about the role and metabolism of DMSP in phytoplankton, it is difficult to know whether their biology is consistent with this proposed mechanism. Whatever the role of DMSP, it is highly variable among species (which means there is no correlation between measures of overall plankton abundance and DMS), and intracellular concentrations vary over five orders of magnitude among species! Additionally, intracellular concentrations are only an indirect control over water-column DMS concentrations, because DMSP is released from phytoplankton cells when they rupture. Fluxes to the atmosphere further depend on its transformation to DMS in the water column and exchange with the atmosphere. The relationship between climate and seawater DMS concentration is mediated by phytoplankton species composition (which is influenced by the balance of resources and not just solar energy) and by the activity of the entire marine ecosystem, which affects grazing rates and heterotrophic activity.

Resolving these problems requires advances in phytoplankton biology and marine ecosystem science, but in the meantime, scientists have come up with creative ways of observing correlations between overall phytoplankton activity, marine atmospheric chemistry, and cloud/aerosol optical properties affecting climate. Studies reviewed by Ayers and Caine found correlations between water-column DMS and incoming solar radiation, suggesting, or at least permitting, a feedback consistent with the CLAW hypothesis. Studies of the unperturbed sulfur cycle in the Southern Hemisphere, which has less land and fewer humans, showed a seasonal cycle of non-seasalt sulfate (NSS-sulfate) and methanesulfonate (MSA) aerosol. Both of these atmospheric constituents are products of the oxidation of DMS. This finding provides evidence supporting a relationship between seasonal changes in the biological source of DMS and resulting atmospheric concentrations. These seasonal cycles were reflected in sulfur aerosols and atmospheric radiation effects. Despite strong evidence supporting parts of the CLAW hypothesis, additional research continues to yield additional, and fascinating, complexity in the linkage of physiological, ecological, physical, chemical, and climatic processes.

Terrestrial ecosystems also play a significant role in the aerosol cycles of the atmosphere. Volatile organic compounds (VOCs) are among the most common condensation nuclei in the atmosphere. Atmospheric scientists have known for decades about the role of VOCs in the atmosphere, but new instrumentation and measurement technologies in the 1990s and in the early years of this century greatly expanded the number of compounds that could be analyzed, and shed light on this complex subject. Atmospheric VOCs have many sources, and some of the first understood were directly related to human activity, notably fossil fuel use. The emission of VOCs from terrestrial plants was discovered in the 1950s (Sanadze 2004). By early in this century, scientists realized that terrestrial biogenic emissions dominated global sources of VOCs, except in locally polluted atmospheres. Biogenic VOCs (BVOCs) play many roles in the atmosphere and can contribute to tropospheric ozone production, and interact with human-caused pollution. In the presence of adequate reactive nitrogen (nitric oxide, NO) VOCs break down chemically to form ozone and can contribute to the atmospheric ozone cycle, much as methane does (Lerdau, Guenther, and Monson 1997).

Volatile biogenic organic compounds are one of the dominant sources of condensation nuclei. They connect the terrestrial biosphere to one of the most important climate elements of the atmosphere in much the same way that, as the CLAW hypothesis suggests, the marine biosphere is coupled with Earth's energy budget. Organic aerosols may be *primary* (primary organic aerosols, POAs)—composed of particles or compounds of biological origin (such as bacteria, spores, plant wax fragments)—or *secondary* (secondary organic aerosols, SOAs) derived by chemical reaction from biogenic source compounds (Fuzzi et al. 2006). Biomass burning is an important source of BVOCs volatized from plant material through incomplete combustion.

While the sources and effects of organic aerosols are complex, the BVOCs have emerged as potentially important in ecosystem–climate interactions. The single most common BVOC, isoprene (C_5H_8, 2-methyl-1,3-butadiene), is emitted from plants—mainly mosses, ferns, and trees—in association with photosynthesis. The production rate of C_5H_8, though much smaller than the overall rate of photosynthesis, tracks light and temperature in similar ways, increasing as rates of photosynthesis increase. Isoprene is energetically expensive for plants to produce, and its exact function and evolutionary significance remain unclear, though it appears to give plant leaves some protection from high temperatures and may compensate for its high cost of synthesis by reducing the energy costs to the plant that occur when the photosynthetic apparatus is damaged or even killed by heat. Because isoprene production responds directly to climate (light and temperature), its production may change with changing climate. Although most isoprene is oxidized in the atmosphere without producing any compounds that act as SOAs, a modest fraction is converted into SOAs. Since the overall global flux of isoprene is very large, this modest fractional yield can produce significant aerosols, especially over productive vegetated regions. Tropical regions have very high isoprene emission rates, so in regions like the Amazon basin, isoprene may play important chemical and physical roles. In such regions, increasing temperature and light could increase isoprene emissions, leading to more aerosols and potentially reduced incoming radiation and lowering temperatures, as suggested by the CLAW hypothesis, but over land instead of the oceans.

The biology of the terpenes is rather different from that of isoprene. Terpenes are produced and stored in the tissues of many plants and perform a multitude of functions (Gershenzon and Dudareva 2007), including conveying resistance to pests and pathogens, and may serve to communicate with insect mutualists of plants. Terpenes are toxic to many bacteria and fungi, as well as to some insects and other potential herbivores, and are distasteful or harmful to some birds and mammals.

The production and abundance of terpenes is not as directly linked to climate as is that of isoprene, whose function may be tied directly to light and temperature. Nonetheless, emissions of terpenes generally increase with warmer temperatures, as do those of isoprene. Because terpenes are volatile, their vapor pressure increases with temperature, and this is at least one reason for the correlation between temperature and emission, but additional biological mechanisms may be operating over seasonal and longer timescales. In contrast with isoprene, terpenes produce a high fractional yield of SOAs during their chemical decomposition in the atmosphere, so even modest terpene sources can have substantial climate effects.

The climate effects of BVOCs extend well beyond their radiative effect. Aerosol effects have been a focus of research in Amazonia. There, increases in aerosols (mainly from biomass burning) alter the size distribution of water droplets in the atmosphere. Increases in the number of droplets, and decreases in the average size of the droplets, reduce rainfall amounts, and may increase the length of the dry season. However, increases in aerosols increase the diffuse radiation fraction and seem to increase rates of photosynthesis, through mechanisms described earlier. Similar dynamics have been hypothesized for the boreal forest as well, where substantial emissions of terpenes occur. Changes to BVOCs and other organic aerosols can have profound effects on the Earth System, affecting temperature directly, modifying the hydrological cycle and precipitation, and altering the carbon cycle.

The BVOCs also have secondary and interactive chemical effects on climate. For example, oxidation of BVOCs and methane requires and consumes the hydroxyl radical, $OH\cdot$, Thus, increasing methane concentrations can increase the lifetime of BVOCs and BVOC SOAs, increasing their radiative effect, which is to cool the atmosphere. Conversely, increasing BVOCs constitutes a sink for $OH\cdot$, which can prolong the lifetime of methane, increasing radiative forcing and warming the atmosphere. The importance of BVOCs as atmospheric aerosol precursors again emphasizes the interconnected nature of the biosphere and the Earth's physical climate system.

Mineral dust, produced in the terrestrial biosphere and transported in the atmosphere, is another important type of aerosol. Although produced on land, dust influences the entire Earth System in interesting ways (Mahowald et al. 2011). It has an initial effect in the atmosphere, where it tends to cause warming, much like a greenhouse gas. Because of their typical large size, dust particles interact more strongly with longwave (thermal) than with shortwave (visible) radiation and so can significantly increase the downward flux of infrared radiation. Dust also affects terrestrial, marine, and cryospheric systems. Mineral dust is often high in nutrients and on long timescales can be an important source of phosphorus in terrestrial systems (Mahowald et al. 2011). In fact, in some regions the main source of phosphorus may be transport in dust, even though it may take tens to millions of years for sufficient phosphorus to accumulate (Chadwick et al. 1999; Vitousek 2004). In marine systems, dust can be a principal source of iron, required for many biological processes in the ocean, including biological nitrogen fixation. Finally, when dust is deposited on snow or ice, it darkens the surface, which can significantly decrease the local albedo, allowing more absorption of solar energy. Dust deposition on snow can greatly accelerate the rate and amount of melting, with significant impacts on the hydrological cycle (Painter et al. 2010).

Ice cores capture dust, along with many other substances, and the ice core record shows that over long periods of time Earth has experienced dusty phases. Dust tends to be produced during dry and windy periods, when vegetation cover is reduced, soil moisture is on the decline, and soils are more vulnerable to erosion. This process may occur over long periods of time—the glacial period tended to be cold, dry, and dusty—but also on shorter timescales, like the familiar Dust Bowl period in the United States during the 1930s. Dust production is controlled by the state of ecosystems, and dust transport and deposition can affect ecosystems, and is another example of a key aerosol-mediated Earth System process.

Ecosystems have complex and diverse influences on climate. Some of these have opposite effects (carbon storage and albedo), others reinforce each other, such as through multiple traces gas pathways. Many ecosystem feedbacks to climate are biogeochemical, mediated through global trace gas concentrations, while others have regional effects, owing to water vapor, albedo, aerosols, and less uniformly distributed atmospheric constituents. Integrating global and regionally synergistic effects requires understanding ecological processes at a hierarchy of time and space scales. Finding simplicity in complexity, and conversely, identifying where simple models break down and require consideration of the entire system, are two of the great challenges to Earth System science.

5 MODELING THE FUTURE

···

HOW WILL ECOSYSTEMS AND THE CARBON CYCLE CHANGE OVER THE NEXT CENTURY?

IN AN ERA OF DRAMATIC CHANGES TO PLANET EARTH, a grand challenge for ecologists has been to foresee how ecosystems might change. Since the possibility of climate change became evident, ecologists have sought ways to predict how ecosystems will change under climate conditions that have no direct analog in the past or present. An important focus has been an attempt to predict how the distribution of ecosystem types or biomes might change in a future climate. A very important area of research has been modeling the effect of climate change on terrestrial and marine carbon storage, to estimate the role of ecological changes in feedbacks to climate. Some models have tried to predict changes to species ranges and patterns of biodiversity as a way of identifying vulnerable regions and species, in both marine and terrestrial habitats.

An emerging focus area is predicting the likelihood of abrupt change. Historical and paleodata suggest that some systems can change dramatically from one state to another when critical climate thresholds are crossed. For example, the change from forest to savanna may occur as

···

a result of only a small change in rainfall. Fire frequency also exhibits threshold effects, leading to major changes in ecosystem structure or type. Lakes can transition from healthy to toxic conditions abruptly, in response to small increases in nutrient loading.

There is evidence that the likelihood of these types of abrupt transitions may be predictable from data during the time leading up to the transition, even though it is not possible to predict what the new system will look like. The possible existence of thresholds implies that there may be limits to how predictable future changes to ecosystems will be and that we may be able only to identify thresholds where significant change will occur, but not what those changes will be. The limits to predictability of ecosystem change are far from being understood, and the material presented in this chapter reflects the earliest stages of an evolving science. The different ways of viewing the future of ecosystems will be discussed, linking conceptual and numerical models to our understanding of the living systems of the planet described in the previous chapters.

LIMITS TO PREDICTABILITY

> It's tough to make predictions,
> especially about the future.
> —Yogi Berra

We start by reviewing some basic theory and background on modeling and prediction. Predicting the effect of future climate is difficult for a long list of reasons. First, one must compute or imagine the climate of the future, itself a difficult task. Then, ecosystems respond to climate at the species level (and there are millions of species), through competition and other interactions among species, and through complex physiological and behavior mechanisms. Some of this complexity has been described in the previous chapters. However, complex systems like the Earth System have structural and mathematical properties that make prediction intrinsically difficult. One of these properties is referred to as "chaos," and the other, "*computational irreducibility.*" Neither of these terms is extremely descriptive about the nature of the underlying issue, and so each will be explained as general background for considering subsequent specific modeling approaches. Both of these lead to a phenomenon called *limits to predictability.*

CHAOS

Many systems of equations and some real-world phenomena exhibit behavior called chaos, in which the future behavior of a system depends sensitively on its initial state. This means that very subtle differences in initial conditions (say, very small temperature or pressure differences in the atmosphere) can result in very different sequences of weather. Imagine initializing a weather forecast model with the best estimates possible of temperature, humidity, and pressure and then forecasting the weather 5 days in advance. Then, imagine varying the initial meteorology—but very slightly—changing the conditions well within the uncertainty of the measurements (probably a few tenths of a degree or less). When the two forecasts are compared, they don't even resemble each other. The model is *sensitively dependent on initial conditions*. If this process is repeated many times, the result will be many different predictions of the weather, depending on small and random changes to the initial conditions. This phenomenon was first described by Edward Lorenz of MIT (Lorenz 1963) and is sometimes referred to as the "butterfly effect" after the idea that the flapping of a butterfly's wings could change the whole course of future weather. Mathematical models of ecological phenomena exhibit chaos (May 2007), but it is much less clear whether biological systems are chaotic. The climate system is chaotic, so biological variability will reflect chaotic patterns caused by the atmosphere (Wang and Schimel 2003).

Although chaotic behavior is a limitation to prediction, only certain phenomena exhibit chaos. For example, weather exhibits chaos. We can't predict whether or where it will rain many days in advance. We don't know where and when hurricanes will form until they are initiated and are visible. However, we can predict many conditions. For example, we know that day is warmer than night, summer is warmer than winter, and high latitudes are warmer than low. Deserts form in highly predictable places, and the weather in the mountains is significantly different from that in lowlands. These phenomena are referred to as *forced* because they arise not from the internal interactions within the system (the atmosphere) but from the configuration of the planet relative to the sun, topography, and other constraints on the atmosphere's behavior. Prediction is not merely a matter of scale. El Niños, which form over similar large spatial scales, are not very predictable, because they arise from the coupled internal dynamics of the atmosphere and ocean, although once they arise, conditions during the El Niño become more predictable. This is why climate scientists believe we can predict the large-scale response of climate to CO_2 even though we can't predict the weather very far in advance.

Chaotic behavior has another property that allows a type of prediction. For example, whereas individual storms are unpredictable more than a few days in advance, rainfall statistics may be much more predictable. That is, although we can't predict storms in advance, we can confidently predict that there will be many more storms near the coast than in the desert. While we can't predict individual events, we can predict differences in the probability of events in different regions. The limits to predictability for chaotic systems have two dimensions. First, how far in advance can specific events be predicted? Second, what behaviors are forced or controlled by external constraints, and which ones arise from internal dynamics? Only by understanding these variables can we target efforts to predictable phenomena and properly characterize the statistical behavior of inherently unpredictable processes.

COMPUTATIONAL IRREDUCIBILITY

Computational irreducibility is a more recent idea, first introduced by Stephen Wolfram (2002). Wolfram argues that most scientific theories greatly simplify an underlying complex reality and allow prediction without explicitly computing all the intervening steps in a process. For example, photosynthesis can be simulated accurately as a function of forcing by light, even though many physical and biochemical processes underlie this seemingly simple relationship. However, given a certain amount of light, these processes nearly always respond in the same way. The process of photosynthesis is *reducible*, and the computations required to simulate it are much simpler than the underlying reality. However, some processes contain complex intermediate steps. At each step many outcomes may be possible, and they are not strongly constrained by the initial forcing. In such a system, it might be essential to simulate each step to link the initial forcing and the eventual response (Beckage, Gross, and Kauffman 2011). For example, climate impacts on species diversity may be dependent on not only the initial species composition but the particular distribution of genetic variability within each population at that time, which is itself dependent on the history of environmental and evolutionary change up to that point in time. At subsequent times in the future, random events could redirect the system, if, for example, rare genotypes are lost. Such a system would be computationally irreducible, in that no simple theory could shortcut calculating all the intervening steps between initial cause and final outcome.

Niche-structured systems may be computationally irreducible (Beckage et al. 2011). While some niche dimensions are related to the biophysical environment other niche dimensions depend on the other interacting species in the environment. Thus, immigration, extinction, and evolution will alter the niche space available to other species, so that continuously computing the changing species (and even genome) composition of an ecosystem might be essential to understanding how it will respond to changing climate. Such a simulation would not reduce climate effects on species ranges to simple deterministic patterns (Kreft and Jetz 2010) and might not be much simpler than nature itself.

Computationally irreducible systems may exhibit the same sorts of limits to predictability that chaotic systems do. For example, some timescales and not others may be predictable. Strongly forced behavior may well be predictable from simple theory even if internal dynamics are irreducible. Certain types of statistical behavior may be predictable when individual events are not. However, these types of systems are less understood and significantly more difficult to simulate than chaotic systems (which may be represented by a simple mathematical formulation, even when their actual behavior is very complex).

PREDICTORS OF CHANGE

In systems whose actual future behavior cannot be forecast, sometimes the likelihood of changes to behavior can be predicted. As a system approaches a critical threshold of some controlling variable at which its behavior may change abruptly, its behavior may oscillate between behaviors typical of its current state and those of its likely future state (see figure 15). This oscillation gives rise to an increase in variance immediately preceding a transition. Detecting such an early warning signal requires detailed time-series observations of critical variables, and there are examples in ecosystems and elsewhere in the Earth System of this type of premonitory behavior (Scheffer et al. 2001).

In other situations, as a system approaches a threshold, barriers to transition weaken. The constraints

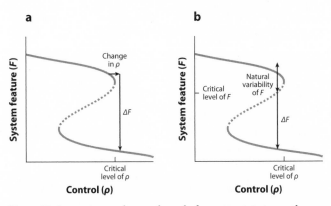

Figure 15. A system can change abruptly from one state to another as a result of either small changes in a control variable (such as climate) or as a consequence of natural variability within the system: (a) an abrupt transition as a result of a change in the control (change in ρ); (b) a change as a result of random variation in a system feature (e.g., vegetation).

(Source: Lenton 2011)

(climate, fire frequency, species composition, or whatever) that held the system in its previous state begin to weaken, giving it access to other states. Prior to the transition, variability in the system is less constrained, and there are fewer forces restoring the system to its initial state. The system may spend more time in states far away from its previous average state. In such a situation, we can examine the temporal behavior of the system and observe a phenomenon called *critical slowing*. The timescale of variability slows as the system approaches a transition and as it spends more time in states similar to but varying around its mean state. Finally, the constraints that previously restored the system to a defined state weaken so much that it transitions to another, adjacent state. The slowing rate of variability is then followed by an abrupt transition to a new state, and entry into a new variability regime.

The preceding two approaches are based on forecasting future ecosystem behavior using observations informed by a theory of change rather than by focusing on detailed mechanistic simulations of ecological or Earth System processes. Sometimes, observing strategies leading to such early warning systems may even be suggested by simulation modeling. Observation-based approaches are critical in systems that are chaotic or irreducible, because while they may not allow long-term forecasts, they provide ongoing information for adaptive responses. The subsequent sections discuss both simulation-based and observation-based approaches to thinking about the future of the world's ecosystems.

CLIMATE MODELS

Projections of future climate from numerical simulation models have been available to ecologists for roughly two decades, and these projections have increased in detail and number. Many models now produce such projections, and simulations correspond to many potential energy, fossil fuel emission, and climate futures (Donner and Large 2008). Before one can anticipate how the biosphere might change in the future, information must be available on future climates. While some ecologists have used highly idealized or conceptual future climate, such as one including a 3° temperature increase, corresponding to global mean changes, for reasons discussed in chapters 1 and 2, most ecologists prefer to use the most detailed projection possible, including extremes of temperature and moisture (droughts and floods), using multiple models to gain a sense of uncertainty. Climate model projections increase in uncertainty with increasingly fine scale, and are most robust at continental-to-global scales, but often, local detail is crucial.

What is a climate model? Climate models simulate the behavior of the atmosphere as a function of energy from the sun; of exchange of matter and energy with the land, ocean, and ice-covered regions of the globe; and as a function of atmospheric composition (water vapor, greenhouse gases, aerosols) (see figure 16). Climate models are based on robust fundamental physical and chemical principles. Nevertheless, many processes are understood imperfectly or cannot be explicitly resolved with today's computing resources. For example, the physics and chemistry of aerosols is still a research area, and individual clouds cannot be resolved. Figure 16 shows the structure of a modern climate model. Models like this can resolve processes down to length scales of tens of kilometers, although these details are only statistically meaningful. Climate models are solved with very fine time steps, on the order of minutes, to resolve fluxes in a physically meaningful way, but again, as with

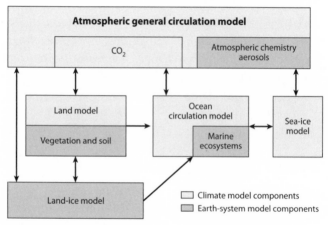

Figure 16. The structure of a modern climate system model. The components, represented by numerical models, are coupled by the fluxes of water, energy, and gases (such as CO_2). These models typically require a million lines of code and run on some of the world's largest supercomputers. The coupled modeling system is represented by simple compartments but represents each component as a geographic grid in which processes are simulated spatially.

(Source: Donner and Large 2008)

spatial scales, averages over longer periods of time are more reliable.

When we use climate models (in coupled or un-coupled modes; see further discussion), we must account for the uncertainty in the physical model as well as the resulting biological uncertainty. Climate model uncertainty has several dimensions. First, future emissions are uncertain because of political, technological, and economic unknowns. Second, the relationship between fossil emissions and atmospheric concentrations is uncertain, because of our incomplete knowledge of land and ocean sources and sinks of greenhouse gases. Third, for any given atmospheric concentration, there is remaining uncertainty in the response of the global climate system (Kiehl, forthcoming). Finally, for any given global mean change, models differ in the distribution of temperature and precipitation changes around the globe even when the predicted global average conditions are similar. In addition, climate models are chaotic (Lorenz 1963), so subtle differences in initial conditions will produce different sequences of weather within the climate model. This property is used in climate change modeling. Multiple *ensemble members* are run with a single climate model, and each model is initialized with slightly different conditions. By averaging over these parallel simulations, robust, forced behavior that occurs across all ensemble members can be distinguished from more stochastic variability. All these physical uncertainties contribute to the eventual computed biological uncertainty.

TERRESTRIAL BIOMES OF THE FUTURE

The repeated patterns of correspondence between climate and ecosystems, and the evolution of similar adaptations in similar terrestrial and marine habitats, are
fundamental facts of ecology. This view of convergence,
while not without significant exceptions and some dis-
pute, leads to the idea that climate effects on vegetation
should be predictable from climate changes and has
given rise to many models. This belief suggests that the
vegetation–climate change is reducible and not chaotic,
at least at large scales—a view to which we'll return in
considering testing and validation of these models.

The earliest biome models, linking climate to ecosystem type (desert, temperate forest, boreal forest,
tundra, etc.) were conceptual and graphical and are usually described as classifications. Common ones include
the Holdridge Life Zone classification and the Koppen
system, although much earlier examples exist. Alexan-
der von Humbolt developed a system in the 1860s, and
Merriam's Life Zone system for describing elevational
changes is still widely used in the western United States.
These systems all assumed that—at least at large scales—
vegetation was at equilibrium with climate and emerged
in a time when climate was largely seen as stationary.
Some of these systems (such as the Koppen) focused on
classifying climate and viewed vegetation as an expression of climate, while others were more focused on mapping vegetation (see figure 5) using climate as correlates.
All these systems focused on generally similar variables,

although they differed in focus on seasonal or annual temperatures, use of linear or nonlinear scales, and incorporation of moisture.

Key variables always include temperature, but water is often at the heart of these models. For example, Holdridge developed a logarithmic scale focusing on annual temperature and moisture conditions. Many systems include both a measure of *water availability*, often simply annual precipitation, and *evaporative demand* (which must be computed according to some sort of model). Including both factors accounts for the relative availability of precipitation-derived water in warmer and colder, or more or less humid, environments. Other things being equal, a higher fraction of precipitation will be available to support plant growth in cooler and more humid environments. That is, to support a given level of plant growth, more precipitation is needed in environments where evaporation is high. Often, the ratio of precipitation to potential evapotranspiration is used to index water stress. Taking into account the central role of evapotranspiration is similar to modeling of carbon uptake in terrestrial systems, or in empirical models of biodiversity, discussed in earlier chapters, and reflects the central role of water and energy availability in supporting and structuring living communities.

Having defined these indexes of climatic resources and stress, biogeographers have used them to describe the spatial distribution of vegetation in a number of ways. The earlier systems focused on vegetation patterns variously called biomes, associations, communities, or ecosystems; other terms are found in the literature as well. Here we use the term *biome*, defined as a complex biotic community with distinctive plant and animal species and persisting stably under the climatic conditions of the region. Using a system based on temperature, precipitation, and potential evapotranspiration, the Holdridge system classifies biomes worldwide into 38 types. The boundaries between these types can be described mathematically and can be incorporated into numerical models (Emanuel, Shugart, and Stevenson 1985). Such models can then be used to study biome responses to future equilibrium changes in climate, or to model paleo-distributions of biomes. For example, Prentice and Fung (1990) used a biome distribution model to calculate the difference between the present and 18,000 years in the past (last glacial maximum) to see whether changes to vegetation contributed to the difference in atmospheric CO_2 between then and the present, and then extrapolated the model to compute the eventual effect of a long-term doubled CO_2 climate. They concluded that changes to terrestrial vegetation could not explain the glacial–interglacial change in atmospheric concentrations.

These models simulate only the equilibrium distribution of vegetation with climate and so can't provide insight into what would happen during a period of dynamic change. If the process of vegetation response to climate is computationally irreducible, then this is a serious flaw, because the final state might not be a deterministic response to climate but, instead, might be a unique response to the events that occurred during the transition. Although early modelers did not explicitly address the issue of irreducibility (indeed, the concept was not yet articulated formally), they did recognize early on that biomes might be too coarse and that future biomes might assemble from species in unique ways, giving rise to *no-analog* biomes, or vegetation patterns that did not occur in the present or in the paleorecord. To address this problem, ecologists and biogeographers sought to define *plant functional types* (PFTs), or groups of plant species that could be considered functionally identical in terms of their response to climate (or some other control). Using this approach, one could model a recognizable biome as the outcome of climatic selection for a mix of PFTs that taken together make up the vegetation of the biome. Thus, for example, a cool-temperature steppe might arise from the dominance of grasses together with shrubs.

Biogeographers and ecologists have tried many schemes for classifying plants into PFTs or general types. These classifications have been based on a wide variety of criteria, from primarily structural properties to physiological and functional properties. Many approaches use a physiognomic approach, in which functional attributes are inferred from visible structural features. Some of these systems are global in extent, but many focus on more limited regions. Most global PFT systems begin with criteria similar to those for biome models. They consider plant tolerances to temperature and moisture. Widely used systems include a range of other variables, such as degree of shade tolerance (distinguishing

over- and understory vegetation or early and late successional species), allocation strategies (to leaves versus
wood or roots), metabolic strategy (C3, C4, nitrogen fix-
ers), longevity (annuals, perennials), and other controls
over plant growth, reproduction, and mortality (Box
1996). These attributes are then linked empirically to environmental constraints to allow mapping of PFTs based
on the ambient environment (climate, perhaps soil properties, etc.). One such scheme (Box 1981, 1996) was based
on 115 PFTs worldwide and was used to identify resulting biomes. That scheme suggested that 15 biome types
are sufficient to describe natural (not agricultural or in-
tensively managed otherwise) vegetation worldwide. For
example, the Box scheme suggests that tropical rainforests occur where the vegetation is dominated by tropical
evergreen broad-leaved trees that have tall woody structures and are evergrowing (able to take advantage of the
lack of seasonality).

This approach attempts to address the problem of irreducibility by posing an alternative model. A review
of the attributes included in the PFT definitions reveals
that they include not only direct impacts of climate (cold
tolerance) but also attributes linked to competitive status (shade tolerance). The Box approach implicitly recognizes that a biome reflects the influence of climate
but also includes species or PFTs whose abundance or
functional role in the biome is determined by their competitive status. For example, a forest that includes shade-
tolerant and shade-intolerant PFTs will have an overall
appearance dominated by the shade-intolerant types but

is very different from a forest that does not also include subordinate shade-tolerant species. In the first case, the forest habitat could be much more complex than the other, with an upper canopy of sun-loving species and a lower canopy of shade-tolerant trees. Simply taking into account the macroclimatic limitations of the trees would not allow identification of the dominant strategy controlling the system's overall structure.

This approach suggests that competitive and successional processes (see figure 17) play out as climate change occurs, and species sort out their presence and abundance based on competitive and facilitative interactions, but the eventual outcome at the level of PFTs (but perhaps not species) is deterministic and predictable. Box's approach recognizes that species interactions occur and are important but suggests that their effects, or eventual

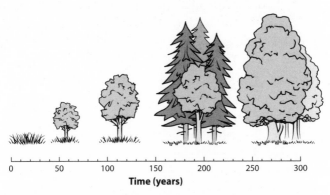

Figure 17. A typical successional sequence, from grasses and bare ground to mature forest as an ecosystem recovers from a disturbance that removed the previous vegetation (a wildfire, for example).

(Source: www.biologycorner.com)

175

effects, are dominated by climate constraints. This implies that the process is reducible, or at least at the level
of gross structural types of vegetation, biome distributions are reducible and that some level of vegetation response to climate change is predictable. This argument is
plausible because of work such as Mooney's on conver-
gent evolution and some aspects of the paleorecord, but
is called into question by unforeseen interactions with
pests and pathogens, as occurred in the Rocky Mountain pine beetle outbreak, described earlier. There, internal dynamic processes changed in ways not predictable
from past climate correlations with vegetation but that
may alter future biome structure and distribution. We
don't know and perhaps can't know whether the general
determinism often found in the paleorecord, or the irreducible and complex changes that some contemporary
observations show, will dominate the future.

Ecologists continued to add detail to vegetation-type
or PFT-based models. The Box-era models linked plant
characteristics to climate descriptively and geographically, and captured interactions between species through
discrete characters like shade tolerance. The next step
in model evolution was to simulate plant growth and
to define PFTs not through discrete characteristics but
through differences in parameter values of mechanistic
models. Models of the Box (1996) type have a primary
limitation when applied predictively: they are calibrated
against the present and the past. Competitive interactions and growth responses, which actually determine
climate responses, are inferred from how these processes

have played out in the past. However, these processes may play out differently in the future. In most cases, calibrated models like the Box model (1996) are successful when conditions don't differ too much from the observations used to create the model. At least in principle, the more mechanistic the model, the better the chance that it will respond correctly to novel conditions outside the calibration range of the data.

Models often called *global vegetation models* (GVMs) or *dynamic global vegetation models* (DGVMs) combine rule-based approaches like the Box (1996) model but add simulations of plant growth and ecosystem processes. Similar models are also being developed for the oceans. Models include simulations of climate controls over photosynthesis, respiration, and other ecosystem processes such as nutrient cycling (Neilson 1995; Bachelet et al. 2001). Each PFT is translated into a set of process parameters. For example, C4 and C3 grasses differ in their maximal photosynthetic rates and water-use efficiency. Deciduous and coniferous trees differ in leaf longevity, as well as other rate constants. Each PFT, in addition to representing a group of taxonomically diverse species, now represents a particular growth strategy in response to environmental constraints. At each time step of the model, each PFT within a model cell grows as a function of modeled photosynthesis, respiration, and nutrient uptake. Depending on the climate, different PFTs will accumulate different amounts of biomass, which allows one to determine which PFTs will succeed in that grid cell and which will disappear. This process is a substitute for actually modeling population dynamics and allows change to occur continuously and with a rate based on plant growth rates. In the previous generation of models, only equilibrium rather than dynamic changes could be simulated.

DGVMs represent a third phase of vegetation models. First, models moved entire biomes around as climate changed (Prentice and Fung 1990; Emanuel, Shugart, and Stevenson 1985). Then, PFT models used similar climate rules to move functional types and build biomes up from functional-type distributions, allowing new biomes to form. DGVMs allowed biomes to shift as PFTs' growth rates increased or declined with changing conditions. More advanced models include interactive effects in the form of competition for water and nutrients, which are themselves influenced by the changing supply and demand for these resources as PFTs and the environment change together. These third-phase models are thus "dynamic" and have time steps to capture the rates of change of biomass at the PFT level (Sitch et al. 2003). Some of these models include other processes, such as fire, by linking fire to climate and vegetation conditions, and simulating the distribution of PFTs between fire-tolerant and fire-intolerant conditions. These models don't simulate all the factors that control vegetation dynamics—processes such as seed mobility and other dispersal mechanisms, barriers to species migration such as urban or arid zones, and previously unknown competitive or mutualistic mechanisms. In short, they add elements of mechanism but don't guarantee that the critical processes that will actually govern biome shifts in an unknown future are included. These models, representing the current state of the art in many models of the coupled Earth System (atmosphere, oceans, and land), are continually being modified and are becoming increasingly complex as additional mechanisms are included.

MARINE ECOSYSTEM FUTURES

Marine ecologists have focused less on biogeographic models, of the types described for terrestrial ecosystems, and more on ecosystem models that couple primary production (photosynthesis) to food webs. The bottom of the marine food chain, phytoplankton, are consumed by zooplankton at extremely high rates (the entire phytoplankton population of the world turns over weekly), so effects of changes in primary productivity are nearly immediate for other trophic levels. At the same time, changes to predators and herbivores, with their high consumption rates, can also have immediate effects on primary productivity. Warming conditions and fresher water at the surface at high latitudes (from melting polar ice and increased precipitation) should reduce overall ocean circulation (Brierley and Kingsford 2009; Doney and Schimel 2007). This change will reduce nutrient flux to the surface and, together with increased stratification from warmer conditions, should reduce overall marine primary productivity. Observations of the world's oceans show clear warming patterns and reductions in marine primary productivity (Behrenfeld et al. 2005), as detected by satellite measurements. The observational time series support the theoretical and model-based prediction that warmer conditions lead to increased stratification, resulting in reduced nutrient availability and hence decreased phytoplankton growth. Since the dominant plants in the oceans live for days instead of decades, as on land, these dynamics are easier to observe, and observations are more straightforward to extrapolate. However, the consequent effects on marine food webs and the production of marine ecosystem services, especially fisheries, is much harder to forecast. We do know, for example, from the El Niño experience, that large decreases in nutrient-fed phytoplankton growth result in declines or migration of fish in higher trophic positions. It is likely that global reductions in marine primary productivity will have significant effects on fisheries and marine biodiversity, but the details are very hard to discern.

Variants on PFT or biome-structured models (in terrestrial and marine systems) use a size-structured representation (Medvigy et al. 2009). These models represent living biomass (for example foliar biomass, woody biomass, and soil carbon on land, or phytoplankton, zooplankton, and fish) as a single number at each location simulated to characterize size distributions within each grid cell (trees or fish of different sizes and ages). On land, this representation makes time since disturbance a critical property, because this period tends to define how old and large individual plants are. This approach adds a great deal of information, since the biological responses and behaviors of organisms often differ with size or age. For example, as trees grow, their water and nutrient demands increase, so size determines how sensitive they are to variation in resource availability. In plants and marine organisms, size or age may determine reproductive status. If age or size distribution (the demographic pyramid) does not vary with time, these effects can be averaged, but if disturbance (fire or disease) or management (forest harvest, fishing pressure) causes the age/size distribution to vary over time, then size/age distribution is critical to understanding the behavior and evolution of the system. Size and age are usually tracked within PFTs or other functional units (fish), so this approach blends naturally with the PFT-based approaches of earlier models. Once size and age are tracked explicitly, even as distributions rather than as individuals, then light and other size-mediated aspects of plant competition can be simulated. Disturbance can be introduced naturally, by changing age and size distributions to reflect mortality and recruitment of new plants after disturbance, allowing successional processes to be simulated far more realistically than in simple PFT models. While this approach still does not directly address migration rates and mobility, it adds incrementally more realism to simulated rates of change. Adding age structures increases model complexity and adds to the complexity of potential behavior while allowing some critical processes (carbon uptake by forests recovering from disturbance, a global phenomenon) to be represented naturally.

CARBON FUTURES

Many models utilized to look at terrestrial carbon under future climate have developed from different rootstock than the DGVM/GVM models. Climatologists and ecologists realized that there were important physical feedbacks between the land surface and the atmosphere and began to build models focused on these interactions (land surface models). Initially, these models focused on evapotranspiration, albedo, and other purely physical (albeit with some biological control) feedbacks. These models, operating as components of climate or Earth System models, are referred to as *coupled* models.

At the same time, ecosystem scientists were interested in carbon and nutrient dynamics in ecosystems at site to regional scales (Pastor and Post 1988; Parton et al. 1987). They developed models that could be used to simulate responses to observed climate patterns, or to climate scenarios derived from climate model projections (Schimel et al. 1991). In these models, plant processes of photosynthesis and respiration are included, and plant mortality, litterfall, microbial decomposition, soil carbon stabilization, and nitrogen cycling are also simulated. These models have been used to analyze the potential effects of climate change (derived from climate model projections) on terrestrial NPP and carbon storage (Melillo et al. 1993). Ecosystem models rarely simulate vegetation dynamics but instead begin with a map of biome or PFT distributions. Each biome (or PFT) has unique parameter vales (as in DGVMs) that govern carbon, water, and nutrient dynamics but generally do not allow for biome boundaries to actually move (Vegetation/Ecosystem Modeling and Analysis Project; VEMAP 1995). Stand-alone ecosystem models are also called *uncoupled*. As Earth System scientists became increasingly interested in carbon–climate feedbacks, more and more aspects of terrestrial carbon cycling were transferred from ecosystem models into coupled models so that interactions with the evolving climate could be simulated.

The coupled/uncoupled model duality also addresses complex behavior in the Earth System. Running a climate model and then using the climate model output as input to an ecosystem model assumes that the climate and the terrestrial carbon system will evolve identically, whether coupled or uncoupled. For this to be true, the coupled carbon–climate system must be neither chaotic not irreducible. In effect, running the two models uncoupled assumes that the systems are reducible, that is, their intermediate steps during change don't ultimately determine the outcome. But complex feedbacks between climate and ecosystems do occur, so fewer and fewer global studies are now carried out with uncoupled models. The extent to which this system is well behaved mathematically, or nonlinear and complex in its responses, remains unclear at the global scale, and this concern underlies much of the anxiety about carbon cycle thresholds or *tipping points*—levels or rates of climate change that could cause the terrestrial system to change dramatically and abruptly.

As coupled models evolved, the ecosystem and DGVM approaches were sometimes merged, so there are DGVMs that include nearly all the processes included in ecosystem models while also simulating vegetation dynamics (Bachelet et al. 2001). Although these models at least potentially exhibit complex behavior, numerous comparisons of models with other models, and models against data, suggest several important generalities about ecosystem response to climate (Schimel, Braswell, and Parton, 1997; Schimel et al. 2000; Gregory et al. 2009). Today, the land biosphere takes up between 1 and 2 Gt of carbon per year, and the oceans take up a similar but less variable amount. In the future, these fluxes could decrease, leading to an increase in the rate of atmospheric concentration increase, or even change sign and release carbon to the atmosphere (Woodwell et al. 1995). When we evaluate the sensitivity of the carbon cycle to climate (and other perturbations), we compare it with these baseline present-day fluxes. Models are calibrated to match observed present-day terrestrial uptake, so they will diverge from present values as a function of climate and other environmental and assumed land-use changes in the future.

In these models, increasing atmospheric CO_2 tends to cause increases in ecosystem carbon storage. Warmer temperatures increase carbon uptake in moist regions but decrease carbon uptake in drier regions. Drier conditions tend to cause carbon loss, while increases in moisture cause enhanced carbon uptake. Temperature effects dominate in wet regions, and moisture dominates in dry regions. However, in regions with intermediate levels of rainfall, warmer conditions may increase evaporative demand and, in effect, cause these areas to become drier. Geographically, many global models show increased carbon uptake in the cooler biomes of the Northern Hemisphere but accelerating carbon loss from tropical ecosystems as they become drier.

Essentially all terrestrial carbon models show *positive* feedback. As temperatures warm, ecosystems take up less carbon (or lose more) and so contribute to increasing atmospheric concentrations. The extent of the increase depends on the model (some are more sensitive to positive CO_2 effects; others are affected more by water or temperature stress). The impact of climate also depends on the climate simulated in the physical model. Some climate models project more extreme responses of temperature or larger changes to precipitation than others. It can be difficult to separate out these effects (of the climate model versus the carbon component) in a coupled model (see figure 18). Models used to compute carbon–climate feedbacks are enormously complex (more than a million lines of computer code) and expensive to operate. Can we find ways of indexing model behavior so we can make qualitative sense of models and compare them to identify critical differences?

For climate models, we can define the *climate sensitivity* (Cess et al. 1989). The constant λ is the climate feedback parameter. The equilibrium climate sensitivity $T_{(2 \times CO_2)}$ is defined as the change in steady-state temperature T for a doubling of the atmospheric CO_2

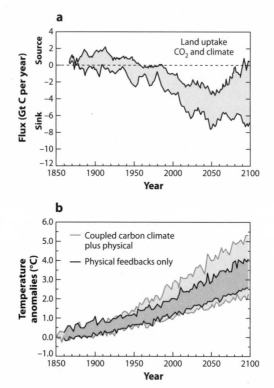

Figure 18. Differences among models of the carbon cycle feedback roughly double the uncertainty in projections of future temperature, mainly because of terrestrial feedbacks to atmospheric CO_2. (a) Range of differences from terrestrial carbon flux changes predicted by Earth System models in simulations of climate change. (b) Range of intermodel projections of future temperature, showing the increase in uncertainty from physical-only to coupled carbon–climate models. (Source: Friedlingstein et al. 2006)

concentration; if the radiative forcing (the amount of additional heat trapped in the atmosphere) resulting from doubled CO_2 is $F_{(2 \times CO_2)}$, then

$$T_{(2 \times CO_2)} = \frac{F_{(2 \times CO_2)}}{\lambda}$$

The feedback parameter, λ, captures the additional feedbacks that might amplify or diminish the effect of CO_2, including the water vapor feedback, effects of clouds, and other physical processes. Can we estimate the sensitivity of the carbon cycle in a similar, simplified way? Pierre Friedlingstein and colleagues (2006) developed a simplified expression for ecological carbon cycle feedbacks. The change in the carbon cycle over time can be (simply) described by the equation

$$C_E = C + C_l + C_o$$

where C_E is the total carbon emitted from fossil fuels and deforestation, C is the change in atmospheric concentration, C_l is the change in land carbon, and C_o is the change in oceanic carbon. If we aggregate land and ocean carbon for the feedback analysis, then

$$C_E = C + C_\beta + C_\gamma$$

where C_β is the change in carbon storage due to the direct effect of CO_2 concentration changes (in land and ocean), and C_γ is the effect of climate on carbon storage, through changes to temperature, precipitation, seasonality, and so forth. Then, we can write

$$C_E = C + \beta C + \gamma T$$

Thus, the net effect of emissions on concentration results from the combined effects of changing CO_2 concentrations and climate. The effect of increasing CO_2 concentration is, generally, to increase carbon storage, and the effect of climate change is to decrease carbon storage. The coefficient β represents a negative feedback (reduces warming), whereas γ is a positive feedback. These parameters can be estimated for complex models (Friedlingstein et al. 2008), and the effects of physical feedbacks to climate (λ) can be separated from biogeochemical feedbacks (β, γ). When this is done, the net effect of carbon cycle feedbacks can be described qualitatively and in the abstract, that is, we can gain a sense of how these effects control future climates in any given model, and we can examine how these parameters vary among models. This analysis can allow us to focus on why these models differ and what effects different model structures have on future climates, and why. In all current models, the net effect of the two carbon cycle feedbacks is positive (they increase warming for any given fossil fuel emission scenario). The net effect of carbon–climate feedbacks is as large as the physical feedbacks (Gregory et al. 2009), and their poor quantification (large differences among models) contributes as much uncertainty to projections of future climates as does physical uncertainty.

The global-scale patterns of extremely complex models can be reduced to a simple set of equations. Although this type of simplified modeling provides useful information for certain questions, it is not relevant for others. For example, knowing the relationship between emissions of fossil fuel carbon (and from deforestation) and atmospheric concentrations is essential for policy. If global policy is eventually decided on that limits climate change to a certain level (probably indexed to global mean temperature), this temperature will correspond to a certain level of greenhouse gas concentration. Emission levels will have to be set globally that produce these concentrations, but these concentrations depend not only on emissions but also on the coefficients λ, β, and γ. The value of λ determines the actual climate change that occurs at a given concentration, and β and γ determine the size of the carbon cycle feedback. If the feedbacks are large and positive, then emission reductions will have to larger than if γ is small (meaning a smaller positive feedback). The global carbon cycle may be deterministic at the global scale, and simplified models may be adequate for developing global policy.

However, these simplified models may also be inadequate. As described earlier, the net effect of climate on carbon depends on the details of temperature and precipitation changes in different regions of the world. This means that fairly subtle spatial details of simulated climate and vegetation may influence the size of the climate feedback, in a way that cannot be computed without considering a great deal of spatial detail. The actual sensitivity of the vegetation to climate may depend on phenology, changes to PFTs, or even species distributions. Therefore, it is always important to have both simple and complex models and to be able to understand the sensitivities of the different mathematical formulations. For example, if we know that estimates of λ are highly sensitive to species distributions, we can estimate uncertainty in projections appropriately and also focus observational and experimental research on those questions. This type of modeling exemplifies reducibility, or simplifying model behavior by leaving out many of the intermediate computations while still reaching similar results.

Model intercomparisons have been essential for learning about the carbon cycle. By comparing models and discovering what behaviors they have in common, we can gain some confidence in the likelihood of those outcomes. By identifying differences among model behavior, we can discover areas of scientific uncertainty and begin to understand how different model structures lead to different outcomes. Similarities and differences among models can then be used to develop critical tests of model predictions against observations and experiments. In effect, we use the models to generate hypotheses, and seek data to test these hypotheses. Model-data comparisons must be interpreted with sophistication. For example, many modelers compare their simulations to features of the carbon cycle such as the seasonal and diurnal cycles, assuming that these test model responses to light and temperature (which vary seasonally). However, in some intercomparisons, many models with extremely different mathematical formulations were able to simulate these cycles about equally well. Does this imply that all the models were correct? Unlikely, since they were based on very different assumptions. Instead, this result suggests that diurnal and seasonal behavior can be simulated based on multiple hypotheses about controls over the carbon cycle, and these data are insufficient to test and possibly reject the fallacious theories.

Even when spatial detail and model complexity don't add to the accuracy of predictions, they can make models more useful and more easily compared with data collected at local scales. The goal of limiting climate change is to limit damage to ecosystem services and human systems. But if we reduce a global model to a simple set of equations with a few parameters describing global average behavior, the model will contain no information about which ecosystems are affected by climate; what the effects, beyond the average carbon storage change are; or how those changes might be perceived by human communities. To understand the consequences of environmental change, we need a considerable amount of detail about where, when, and what changes will occur. Complex models are required to provide this level of biological detail. The multiple demands on Earth System models lead most scientists to conclude that there is no one level or type of model that can serve all needs.

Complex models provide local detail useful in some contexts, while simple models condense the complexity to a few principles to guide our understanding. One compromise is to downscale, or to infer detailed patterns of change from coarse-resolution models. This can be done by using statistical relationships, by coupling coarse- and fine-resolution models and various other techniques to aid in predicting change robustly in global models while inferring consequences in a more detailed fashion (Wilby and Wigley 1997). From the interplay of complex and simple models we can begin to gain a sense of what features of the carbon cycle and global ecosystems are predictable, and what the limits are to prediction.

6 SUMMARY AND CONCLUDING THOUGHTS

....

THE INTERACTION OF CLIMATE AND ECOSYSTEMS HAS always affected human society: consider the defining nature of the Dust Bowl in U.S. history or the impacts of recent events such as Hurricane Katrina or the heat-wave deaths in Europe in 2003. Climate changes caused by fossil fuel burning are already affecting humanity and the resources on which we depend. Preparing for increasingly certain environmental changes in the future will demand responses from scientific, engineering, economic, and political sectors of society, but always within the context of our understanding of the Earth System.

Understanding climate and ecosystems in the Earth System continues to challenge researchers, and the complexity of this system makes communicating even its well-known attributes to civil society a challenge. The sheer number of components, interactions, and timescales means that researchers must use models to understand and communicate about Earth System behavior, but its complexity and geographic diversity means that theoretical principles are not sufficient; extensive observations are also required to describe and predict its behavior.

....

Neither fully disciplinary, nor interdisciplinary, Earth System Science focuses on processes that integrate the planet's spheres. For example, research on the carbon cycle requires theory and observations about specific socioeconomic, atmospheric, oceanic, and ecological phenomena, but much of the science in those disciplines is only tangentially relevant to carbon cycling. However, understanding the carbon cycle requires depth in those processes that directly affect carbon dynamics, in whichever sphere they may occur. Study of the carbon cycle—or, more broadly, climate and life—is a coherent intellectual activity that requires both breadth and depth. Many of the problems that face humanity today, and their solutions, require this same combination of broad and deep thought.

By contrast, early analyses of the terrestrial and marine biospheres usually focused on specific processes. Some focused on biogeography, some only on biophysics, and others only on primary production and biogeochemistry. Some biogeochemical models include only idealized functional types, while others explicitly include trophic processes and population dynamics. More recent models address species ranges or patterns of biodiversity. Each of these approaches is useful in addressing certain questions, timescales, and spatial extents. However, many of the important processes emerge when physical and biogeochemical processes interact, when population and evolutionary changes affect physics, or when long-term processes affect the statistics of extremes. We don't yet understand the Earth System and its interactions well enough to know when the answer to a given question lies in simplifying the problem, and when the answer emerges from the full complexity of the system.

Society now depends on weather forecasting, and life-and-death decisions routinely hinge on the skill of the forecast. Forecasters not only understand how well they can forecast on average but have a clear knowledge of those phenomena that are easier or more difficult to predict. Even after decades of regular weather forecasts, there is still debate and much room for improvement in how forecasts are communicated, understood, and acted on.

Communicating climate is an even more difficult challenge. Which aspects of change in the biosphere are predictable with simple theory, such as the economic analogy or the Farquhar model? What aspects of the carbon–climate system can be captured by λ, β, and γ, and which require enormously complex models? Which processes in the Earth System are chaotic or computationally irreducible, and which ones are strongly forced and describable with simple theory? Much of the debate over climate science, in both atmospheric science and ecology, revolves around these two paths. Theoreticians may abhor complex and hard-to-understand numerical models with their enormous computing costs, huge data sets, and potential for error. Conversely, modelers see theory as oversimplifying and—although beautiful—failing to predict complex and subtle interactions. Biologists have debated the role of species in models, which addresses the reducible–irreducible argument in a particular.

Complexity, including chaotic dynamics, irreducibility, and thresholds, leads to different approaches to prediction than in systems with simpler behavior. Efforts to predict the behavior of the physical system have taught important lessons. For example, even complex and irreducible systems may give rise to quasi-predictable behavior, and the complexity of their responses may become organized into modes, like the El Niño pattern in the atmosphere. These modes may allow for reducible theories about certain scales in time and space.

We can learn from prediction by using systems that have complex and contingent behavior; in effect, we learn from experience. Figure 19 shows the increase in *skill*, or ability to predict defined metrics, of the National Oceanic and Atmospheric Administration (NOAA) numerical weather prediction (NWP), which uses a numerical model of the atmosphere based on Newtonian physics and includes moist processes (clouds, water vapor, precipitation) and radiation. Forecast models use a technique called *data assimilation*. Weather prediction is an *initial-value* problem, that is, the evolution of the system is predicted as a function of initial state-variable conditions (temperature, pressure, humidity) and model equations. The system is chaotic, so a perfect prediction requires perfect knowledge of initial conditions, which is impossible. Instead, the model predictions of state variables and observations are adjusted to one another using mathematical optimization techniques, to create

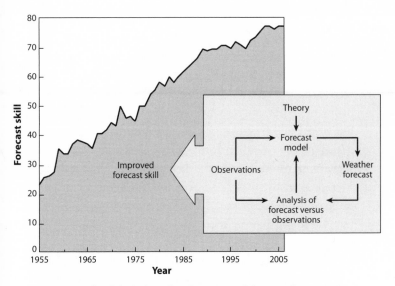

Figure 19. The ability of weather forecast models to predict variation in atmospheric pressure (forecast skill) improved over time, driven by the regular analysis of weather forecasts against observations. The inset figure shows the cycle of observation, assimilation of observations into a model, analysis of results, and comparison with new observations, followed by a new forecast.

the best estimate consistent with the data and the model equations. The model is integrated forward in time, then compared again with observations and again optimized. The iteration process thus minimizes the impact of initial conditions on the forecast. NWP models were initially developed in the 1950s, using some of the world's first computers, and have been continually improved since then. Their skill is monitored operationally and has steadily improved (Randall 2012).

The improvement in skill over 55 years is nearly monotonic and reflects not only improvements in technology, such as satellites, supercomputers, and advanced mathematical techniques, but also millions of encounters of theory with observations via simulation. This improvement not only advances scientific understanding but also leads to more useful information for decision making. The weather forecast is compared with observations multiple times a day, at myriad grid points. Meteorologists are able to evaluate the behavior of the forecast under constantly changing conditions, in effect, sampling an ever-increasing part of the whole set of possibilities in the climate system. From this experience over time, model errors can be used to detect areas for model improvement, and phenomena can be classified by how far ahead they can be predicted. Although the average skill has improved, it has improved much more in certain situations than in others. Forecasts in winter tend to be better than in summer (winter weather is less convective and, statistically, occurs over larger scales), among other patterns of predictability. Without a huge number of model predictions and analyses, phenomena can be classified only in principle, and complex systems theory tells us this can be done only so well without observations.

Biologists have not focused effort on predicting specific sequences of events as much as meteorologists but have emphasized what-if experimental studies, to predict what will happen under certain circumstances. This conditional knowledge has then been built into various types of models that allow decisions to be supported by what-if analyses. In the era of global change, what-if modeling must be complemented by predictive modeling. Forecast modeling is very different from the more monolithic hypothesis testing familiar to biology, in which theories produce predictions that can be tested in incisive experiments. Earth System behavior is sufficiently complex that the fundamental theory leads to chaotic and irreducible models. Today's coupled carbon–climate models resemble weather models of the 1950s and 60s. Coupled models include a core of well-understood theory, but many phenomena are not included or are represented by rough approximations. Just as weather modelers improved NWP models over time by making predictions and analyzing where and when they failed, so carbon modelers are beginning to organize and make systematic model–data comparisons to evaluate and improve their models.

As weather models developed, they often failed in ways that identified data gaps or key unobserved parameters that then helped guide the prioritization of new observations and even new observing technologies. Carbon models can also be used in this way, so that model–data comparisons not only improve models but over time lead to improved observing systems (Baker et al. 2008). In the example described earlier, many models simulated diurnal and seasonal cycles roughly equally well but varied in the realism of other variables, such as biomass, nutrient content, or phenology. Those variables could then be compared directly with observations to distinguish among models.

Theories about ecosystem carbon cycling are likely to be contingent: that is, they predict a certain outcome under some conditions but potentially a quite different outcome under others. For example, warming might increase carbon storage unless it leads to drought mortality or increased insect damage, in which case it will lead to reduced uptake. Regular forecasts and routine comparisons of predictions with new observations (see figure 19) build a database of experience that can help identify simple probabilistic models and identify emergent modes of behavior. Doing so requires building models, collecting observations, making predictions, and comparing them over time. Understanding and managing complex systems requires a data-intensive science different from the theory-centric science that many of us learned. The Earth System approach is essential for increasing knowledge about the complex coupled physical–biological–social processes than govern our planet. Even more important, it is necessary to understand the complex and contingent behavior of the Earth System to inform the management and governance of the planet.

Glossary

Aerosols—Suspensions of fine solid particles or liquid droplets in a gas, such as water droplets or dust particles in the atmosphere.

Albedo—The ratio of reflected radiation from the surface to radiation incident upon the surface.

Anaerobic—The absence of oxygen in an environment, or metabolic processes in that environment.

Biome—A climatically and geographically defined living community of plants, animals, and microorganisms; ecological systems with similar characteristics covering significant areas of the planet as a result of somewhat similar climate, soils, and topography on land, or climate, chemistry, and circulation or mixing in the oceans.

Biosphere—The sum of all living organisms on the planet, and the processes they mediate (such as photosynthesis).

Boundary layer—In fluid dynamics, the layer of fluid adjacent to a surface where the effects of viscosity are significant. The planetary boundary layer of the atmosphere is the lowest portion of the atmosphere where its behavior is directly affected by interactions with the land or ocean surface.

Carbon cycle—The exchange of carbon among all its reservoirs and forms, including the atmosphere, oceans, vegetation, soils, carbonaceous minerals, and rocks.

Chapman reactions—The photochemical reactions that lead to the formation of ozone in the stratosphere.

Chaos—In Earth System science, a description of systems that are highly sensitive to initial conditions, so that small differences in initial conditions produce exponentially diverging outcomes, creating limits to predictability.

Climate—The weather conditions prevailing in an area in general or over a long period, or the long-term average weather.

Climate sensitivity—The quantitative impact of increasing greenhouse gases on climate, usually defined operationally as the equilibrium response of global mean temperature to a doubling of the preindustrial level of atmospheric CO_2.

Chlorophyll—The most common molecule directly responsible for capturing the energy of sunlight in photosynthetic organisms.

Community—A set of two or more different species occupying the same geographical area.

Decomposition—The breakdown of nonliving organic matter to simpler organic compounds and ultimately carbon dioxide, usually mediated by bacteria, fungi, and other microorganisms.

Earth System—The interactive system of *spheres* on planet Earth, including the atmosphere, oceans, cryosphere, lithosphere, and biosphere.

Ecosystem—A community of living organisms (plants, animals, and microbes) together with the abiotic components of their environment (such as the climate, water, and soil), interacting as a system.

ENSO (El Niño/La Niña–Southern Oscillation)—An irregular oscillation of climate patterns that occurs across the tropical Pacific Ocean roughly every five years. The ENSO cycle affects weather globally, even though it is caused by changes in the Pacific.

Eddy covariance—A technique for measuring the exchange of matter and energy at Earth's surface directly, over areas of a square kilometer or so, that takes advantage of transport by small atmospheric eddies just above the vegetated canopy.

Euphotic zone—The sunlit upper layer of an ocean or lake.

Evapotranspiration—The sum of evaporation and transpiration (or plant water use) from terrestrial regions.

Forecast—To estimate or predict in advance. In environmental science, it usually means to estimate the most likely future state of a system based on (1) its condition today (as in the weather forecast) or (2) a decision made today (for example, to forecast the outcome of a forest harvesting regime or a global climate treaty target).

Geoengineering—The intentional modification of climate by manipulating either Earth's carbon system or its radiative energy balance (e.g., by changing albedo).

Habitat—The place or type of place in which an organism lives.

Half-life—The amount of time required for a quantity to fall to half the value initially measured. Often used for radiocarbon or atmospheric gas concentrations.

Latent heat—The heat released or absorbed by a body during a process that occurs without a change in temperature, such as the condensation or evaporation of water.

Methane—CH_4, a key greenhouse gas.

Missing sink—The additional sink, thought to be a terrestrial ecosystem sink, in the modern-day global carbon budget after fossil emissions, atmospheric accumulation, and ocean uptake—which are well constrained by observations—and an estimate of deforestation are accounted for. It is not constrained by direct observation.

Net radiation—The difference between the sum of incoming and outgoing short- and longwave radiation that defines the amount of solar radiation available.

Niche—The biotic or environmental conditions and resources, such as temperature tolerance and food needs, that define the requirements of an individual organism or a species.

Nitric oxide—NO, a highly reactive atmospheric gas.

Nitrous oxide—N_2O, a long-lived greenhouse gas.

Photosynthesis—The process used by plants and other organisms to convert energy captured from the sun into chemical energy used to fuel the organism's metabolism.

Phytoplankton—The microscopic photosynthetic component of the plankton community, occurring in fresh- and saltwater.

Redfield ratio—The atomic ratio of C, N, and P in plankton and throughout the deep oceans. The term is sometimes generalized to mean characteristic stoichiometric element rations in organisms and ecosystems generally.

Regime—A characteristic pattern or behavior of a system that is maintained by interacting and mutually reinforced processes. Complex systems may produce simpler and more easily understood regimes, as occurs, for example, in the El Niño–La Niño system.

Respiration—The processes in an organism in which oxygen or other electron acceptors are conveyed to tissues and cells to support the oxidation of organic matter to provide metabolic energy, resulting in the release of carbon dioxide. The processes in plants, animals, and microbes whereby photosyn-thetically fixed carbon is returned to the atmosphere.

Species range—The geographic area within which a given species occurs, defined geographically or in terms of limit-ing conditions such as temperature, topography, or another organism. For example, a bird's range may be controlled by a plant on which it feeds.

Stoichiometry—Characteristic elemental ratios in com-pounds, organisms, or other components of ecosystems (e.g., soils or water).

Stomate—A leaf pore. CO_2 for photosynthesis and water for transpiration are exchanged through leaf stomates. Stomatal aperture varies with the condition of the plant, allowing regu-lation of water and carbon exchange by the vegetation.

Timescale—A sequence of events over time used as a mea-sure of duration, such as the geologic timescale or the glacial–interglacial timescale.

Trait—A distinguishing characteristic of a species.

Transpiration—The transport of water from the roots of a plant to the stomates, or leaf pores, driven by the evaporation of water from the stomates.

Water cycle—The exchange of water among all its reservoirs and forms, including the atmosphere, oceans, lakes and rivers, soil moisture, and groundwater.

Weather—The state of the atmosphere at a place and time as defined by heat, cloudiness, dryness, sunshine, wind, rain, as well as short-term variations in air quality.

References

THE REFERENCES INDICATED IN BOLDFACE TYPE ARE useful background reading and provide a survey of the work in Earth System Science during the past decades. These papers occur in the literatures of so many disciplines that they can be hard to discover from a starting point in any one discipline. This is my personal best-of list, and these articles are good entry points to the extensive primary literature. Some of these papers are great syntheses; others are early discoveries that led to a large body of literature. The more recent literature is cited in the text, but the highlighted papers are where some of the big ideas first emerged.

Aber, J. D., and J. M. Melillo. 1982. Nitrogen immobilization in decaying hardwood leaf litter as a function of initial nitrogen and lignin content. *Canadian Journal of Botany* 60:2263–2269.

Archer, D. 2010. *The global carbon cycle*. Princeton, NJ: Princeton University Press.

Aufdenkampe, A. K., E. Mayorga, P. A. Raymond, J. M. Melack, S. C. Doney, S. R. Alin, R. E. Aalto, and K. Yoo. 2011. Riverine coupling of biogeochemical cycles between land, oceans and atmosphere. *Frontiers in Ecology and the Environment* 9:53–60.

Ayers, G. P., and J. M. Cainey. 2007. The CLAW hypothesis: A review of the major developments. *Environmental Chemistry* 4:366.

Bachelet, D., R. P. Neilson, J. M. Lenihan, and R. J. Drapek. 2001. Climate change effects on vegetation distribution and carbon budget in the U.S. ecosystems. 4:164–185.

Baker, D. F., H. Bösch, S. C. Doney, and D. S. Schimel. 2008. Carbon source/sink information provided by column CO_2 measurements from the Orbiting Carbon Observatory. *Atmospheric Chemistry and Physics Discussions* 8:20051–20112.

Baldocchi, D. D. 2003. Assessing ecosystem carbon balance: Problems and prospects of the eddy covariance technique. *Global Change Biology* 9:479–492.

Barnola, J., P. Pimienta, D. Raynaud, and Y. Korotkevich. 1991. CO_2–climate relationship as deduced from the Vostok ice core: A reexamination based on new measurements and on a reevaluation of the air dating. *Tellus Series B: Chemical and Physical Meteorology* 43:83–90.

Beckage, B., L. Gross, and S. Kauffman. 2011. The limits to prediction in ecological systems. *Ecosphere* 2:125, doi:10.1890/ES11-00211.1.

Behrenfeld, M. J., E. Boss, D. A. Siegel, and D. M. Shea. 2005. Carbon-based ocean productivity and phytoplankton physiology from space. *Global Biogeochemical Cycles* 19:GB1006, doi:10.1029/2004GB002299.

Bloom, A., F. S, Chapin, and H. A. Mooney. 1985. Resource limitation in plants: An economic analogy. *Annual Review of Ecology and Systematics* 16:363–393.

Bonan, G. B. 2008. *Ecological climatology: Concepts and applications*, 2nd ed. Cambridge: Cambridge University Press.

Bonan, G. B., and L. Sirois. 1992. Air temperature, tree growth, and the northern and southern range limits to *Picea mariana*. *Journal of Vegetation Science* 3:495–506.

Box, E. O. 1996. Plant functional types and climate at the global scale. *Journal of Vegetation Science* 7:309–320.

Brierley, Andrew S., and Michael J. Kingsford. 2009. Impacts of climate change on marine organisms and ecosystems. *Current Biology* 19:R602–614, http://dx.doi.org/10.1016/j.cub.2009.05.046.

Cess, R. D., G. L. Potter, J. P. Blanchet, G. J. Boer, S. J. Ghan, J. T. Kiehl, T.-H. Le, et al. 1989. Interpretation of cloud–climate feedback produced by 14 general circulation models. *Science* 245:513–516.

Chadwick, O. A., L. A. Derry, P. M. Vitousek, B. M Huebert, and L. O. Hedin. 1999. Changing sources of nutrients during four million years of ecosystem development. *Nature* 397:491–497.

Chapman, S. A theory of upper-atmosphere ozone. *Memoirs of the Royal Meteorological Society* 3:103–125.

Charlson, R. J., J. E. Lovelock, M. O. Andreae, and S. G. Warren. 1987. Oceanic phytoplankton, atmospheric sulphur, cloud albedo and climate. *Nature* 326 (6114): 655–661.

Cleveland, C. C., A. R. Townsend, D. S. Schimel, H. Fisher, R. W. Howarth, L. O. Hedin, S. S. Perakis, et al. 1999. Global patterns of terrestrial biological nitrogen (N_2) fixation in natural ecosystems. *Global Biogeochemical Cycles* 13:623–645.

Cody M. L., and H. A. Mooney. 1978. Convergence versus nonconvergence in Mediterranean-climate ecosystems. *Annual Review of Ecology and Systematics* 9:265–321.

Cohen Y., and J. Pastor. 1991. The responses of a forest model to serial correlations of global warming. *Ecology* 7:1161–1165.

Cole, C. V., and R. D. Heil. 1981.Phosphorus effects on terrestrial nitrogen cycling. In Terrestrial nitrogen cycles, processes, ecosystem strategies and management impacts, ed. F. E. Clark and T. Rosswall. *Ecological Bulletin* (Stockholm) 33: 363–374.

Davis M. B. 1986. Climatic instability, time lags, and community disequilibrium. Chap. 16 in *Community ecology*, ed. J. M. Diamond and T. J. Case, 269–284. New York: Harper & Row.

Dickinson, R. E., and R. J. Cicerone. **1986**. Future global warming from atmospheric trace gases. *Nature*, 319:109–115.

Doney, S. C., W. M. Balch, V. J. Fabry, and R. A. Feely. 2009: Ocean acidification: A critical emerging problem for the ocean sciences. *Oceanography* 22 (4): 16–25.

Doney, S. C., M. Ruckelshaus, J. E. Duffy, J. P. Barry, F. Chan, C. A. English, H. M. Galindo, et al. 2012. Climate change impacts on marine ecosystems. *Annual Review of Marine Science* 4:11–37.

Doney, Scott C., and David S. Schimel. 2007. Carbon and climate system coupling on timescales from the Precambrian to the Anthropocene. *Annual Review of Environment and Resources* 32:31–66.

Donner L., and W. G. Large. 2008. Climate modeling. *Annual Review of Environment and Resources* 33:1–17.

Elsig, J., J. Schmitt, D. Leuenberger, R. Schneider, M. Eyer, M. Leuenberger, F. Joos, H. Fischer, and T. F. Stocker. 2009. Stable isotope constraints on Holocene carbon cycle changes from an Antarctic ice core. *Nature* 461:507–510.

Emanuel, W. R., H. H. Shugart, and M. P. Stevenson. 1985. Climate change and the broad-scale distribution of terrestrial ecosystem complexes. *Climatic Change* 7:9–43.

Enting, I., T. M. L. Wigley, and M. Heimann. 1994. Future emissions and concentrations of carbon dioxide: Key ocean/atmosphere/land analyses. CSIRO Division of Atmospheric Research Technical Paper no. 31. Canberra, Australia: Commonwealth Scientific and Industrial Research Organisation.

ESSC. 1988. Earth System sciences: A closer view. Washington, DC: NASA, Earth System Sciences Committee (the *Bretherton* report).

Farquhar, G.D., S. von Caemmerer, and J. A. Berry. 1980. A biochemical model of photosynthetic CO_2 assimilation. *Planta* 149:78–90.

Field, C. 1983. Allocating leaf nitrogen for the maximization of carbon gain: Leaf age as a control on the allocation program. *Oecologia* 56:341–347.

Fine, M., and D. Tchnerov. 2007. Scleractinian coral species survive and recover from decalcification. *Science* 30:1811.

Friedlingstein, P., P. Cox, R. Betts, L. Bopp, W. von Bloh, V. Brovkin, P. Cadule, et al. 2006. Climate–carbon cycle feedback analysis: Results from the C4MIP model intercomparison. *Journal of Climate* 19:3337–3353.

Fung, I. Y., S. C. Doney, K. Lindsay, and J. John. 2005. Evolution of carbon sinks in a changing climate. *Proceedings of the National Academy of Sciences of the United States of America* 102:11201–11206.

Fuzzi, S., M. O. Andreae, B. J. Huebert, M. Kulmala, T. C. Bond, M. Boy, S. J. Doherty, et al. 2006. Critical assessment of the current state of scientific knowledge, terminology, and research needs concerning the role of organic aerosols in the atmosphere, climate, and global change. *Atmospheric Chemistry and Physics* 6:2017–2038.

Gaylord B., and S. D. Gaines. 2000. Temperature or transport? Range limits in marine species mediated solely by flow. *American Naturalist* 155:769–789.

Gershenzon, J., and N. Dudareva. 2007. The function of terpene natural products in the natural world. *Nature Chemical Biology* 3:408–414.

Gregory, J. M., C. D. Jones, P. Cadule, and P. Friedlingstein. 2009: Quantifying carbon cycle feedbacks. *Journal of Climate* 22:5232–5250.

Gulledge, J., A. P. Doyle, and J. P. Schimel. 1997. Different NH_4^+-inhibition patterns of soil CH_4 consumption: A result of distinct CH_4-oxidizer populations across sites? *Soil Biology and Biochemistry* 29:13–21.

Harriss, R. C., D. I. Sebacher, and F. Day. 1982. Methane flux in the Great Dismal Swamp. *Nature* 297:673–674.

Holdridge, L. R. 1947. Determination of world plant formations from simple climatic data. *Science* 105:367–368

Holt, R. 2009. Colloquium papers: Bringing the Hutchinsonian niche into the 21st century; Ecological and evolutionary perspectives. *PNAS* 106, Supplement 2: 19659–19665.

Houghton, R. A. 2003. Revised estimates of the annual net flux of carbon to the atmosphere from changes in land use and land management 1850–2000. *Tellus* 55B:378–390.

Hurrell, J. W., and C. Deser. 2009: North Atlantic climate variability: The role of the North Atlantic Oscillation. *Journal of Marine Systems* 78:28–41.

Jenny, Hans. 1980. *The soil resource: Origin and behaviour*. New York: Springer-Verlag.

C. D. Keeling. 1998. Rewards and penalties of monitoring the earth. *Annual Review of Energy and the Environment* 23:25–82.

Keller, M., M. E. Mitre, and R. F. Stallard. 1990. Consumption of atmospheric methane in tropical soils: Effects of agricultural development. *Global Biogeochemical Cycles* 4:21–27.

Kiehl, J. Forthcoming. *Climate sensitivity*. Princeton, NJ: Princeton University Press.

Kreft, H., and W. Jetz. 2010. Global patterns and determinants of vascular plant diversity. *Proceedings of the National Academy of Sciences of the United States of America* 104:5925–5930.

Lenton, T. M. 2011. Early warning of climate tipping points. *Nature Climate Change* 1:201–209.

Lerdau, M., A. Guenther, and R. Monson, 1997: Plant production and emission of volatile organic compounds. *Bioscience* 47:373–383.

Lorenz, Edward N. 1963. Deterministic non-periodic flow. *Journal of the Atmospheric Sciences* 20:130–141.

Lorius, Claude, Jean Jouzel, Dominique Raynaud, J. Hansen, and H. Le Treut. 1990. The ice-core record: Climate sensitivity and future greenhouse warming. *Nature* 347:139–145.

Mahowald, N., D. Ward, S. Kloster, M. Flanner, C. Heald, N. Heavens, P. Hess, J.-F. Lamarque, and P. Chuang, 2011. Aerosol impacts on climate and biogeochemistry. *Annual Reviews of Environment and Resources* 36:45–74.

Manabe, S., and R. Stouffer. **1993**. Century-scale effects of increased atmospheric CO_2 on the ocean–atmosphere system. *Nature* 364:215–217.

Martin, J. H., and S. E. Fitzwater. 1988. Iron-deficiency limits phytoplankton growth in the Northeast Pacific Subarctic. *Nature* 331:341–343.

May, R. 2007. *Stability and complexity in model ecosystems.* Princeton: NJ: Princeton University Press.

Medvigy, D., S. C. Wofsy, J. W. Munger, D. Y. Hollinger, and P. R. Moorcroft. 2009.Mechanistic scaling of ecosystem function and dynamics in space and time: The Ecosystem Demography model version 2. *Journal of Geophysical Research, Biogeosciences* 114(G1).

Melillo, J. M., A. D. McGuire, D. W. Kicklighter, B. Moore III, C. J. Vorosmarty, and A. L. Schloss.1993 Global climate change and terrestrial net primary production. *Nature* 363:234–240.

Merriam, C. H., and L. Steineger. 1890. Results of a biological survey of the San Francisco mountain region and the desert of the Little Colorado, Arizona. *North American Fauna Report* 3. Washington, DC: U.S. Department of Agriculture, Division of Ornithology and Mammalia.

Mooney, H. A., B. G. Drake, R. J. Luxmoore, W. C. Oechel, and L. F. Pitelka. 1991. How will terrestrial ecosystems interact with the changing CO_2 concentration of the atmosphere and anticipated climate change? *BioScience* 41:96–104.

Moore, B., III, and B. H. Braswell. 1994. The lifetime of excess atmospheric carbon dioxide. *Global Biogeochemical Cycles* 8(1): 23–38, doi:10.1029/93GB03392.

Mosier, A., D. S. Schimel, D. Valentine, K. Bronson, and W. J. Parton. 1991. Methane and nitrous oxide fluxes in native, fertilized, and cultivated grasslands. *Nature* 350:330–332.

Neilson, R. P. 1995. A model for predicting continental scale vegetation distribution and water balance. *Ecological Applications* 5:362–385.

Nemani, R. R., and S. W. Running. 1989. Testing a theoretical climate–soil–leaf area hydrologic equilibrium of forests using satellite data and ecosystem simulation. *Agriculture and Forest Meteorology* 44:245–260.

Painter, T. H., J. Deems, J. Belnap, A. Hamlet, C. C. Landry, and B. Udall. 2010. Response of Colorado River runoff to dust radiative forcing in snow. *Proceedings of the National Academy of Sciences of the United States of America*, doi:10.1073/pnas.0913139107.

Parmesan, C. 2006. Ecological and evolutionary responses to recent climate change. *Annual Review of Ecology, Evolution, and Systematics* 37:637–669.

Parton, W. J., D. S. Schimel, C. V. Cole, and D. S. Ojima. 1987. Analysis of factors controlling soil organic levels of grasslands in the Great Plains. *Soil Science Society of America Journal* 51:1173–1179.

Pastor, J., and W. M. Post. 1988. Response of northern forests to CO_2-induced climate change. *Nature* 334:55–58.

Petit, J. R., J. Jouzel, D. Raynaud, N. I. Barkov, J.-M. Barnola, I. Basile, M. Bender, et al. 1999. Climate and atmospheric history of the past 420,000 years from the Vostok ice core, Antarctica. *Nature* 399:429–43

Porter, W., and M. Kearney. 2009. Colloquium papers: Size, shape, and the thermal niche of endotherms. *PNAS* 106: Supplement 2: 19666–19672.

Powell, J. A., and J. A. Logan. 2005. Insect seasonality: Circle map analysis of temperature-driven life cycles. *Theoretical Population Biology* 67:161–179.

Prentice, K. C., and I. Fung. 1990. The sensitivity of terrestrial carbon storage to climate change. *Nature* 346:48–50.

Randall, D. 2012. *Atmospheres, clouds and climate.* Princeton, NJ: Princeton University Press.

Reay, D. S., and D. B. Nedwell. 2004. Methane oxidation in temperate soils: Effects of inorganic N. *Soil Biology and Biochemistry* 36:2059–2065

Redfield, A. C. 1934. On the proportions of organic derivations in sea water and their relation to the composition of plankton, 177–192. In James Johnstone Memorial Volume, ed. R. J. Daniel. Liverpool, UK: University Press of Liverpool.

Rosenzweig, Michael L. 1968. Net primary productivity of terrestrial communities: Prediction from climatological data. *American Naturalist* 102:67–74.

Rotenberg, E., and D. Yakir. 2010. Contribution of semi-arid forests to the climate system. *Science* 22:451–454.

Russell, Lynn M., Philip J. Rasch, Georgina M. Mace, Robert B. Jackson, John Shepherd, Peter Liss, Margaret Leinen, et al. 2012. Ecosystem impacts of geoengineering: A review for developing a science plan. *Ambio*, http://www.springer.com/environment/journal/13280.

Saatchi Sassan S., Nancy L. Harris, Sandra Brown, Michael Lefsky, Edward T. A. Mitchard, William Salas, Brian R. Zutta, et al. 2011. Benchmark map of forest carbon stocks in tropical regions across three continents. *PNAS* 108 (24): 9899–9904, doi: 10.1073/pnas.1019576108.

Salati, E., and P. B. Vose. 1984. Amazon basin: A system in equilibrium. *Science* 225:129–138.

Sanadze, G. A. 2004. Biogenic isoprene (a review). 2004. *Russian Journal of Plant Physiology* 51:729–741.

Scheffer, M., S. Carpenter, J. A. Foley, C. Folke, and B. Walker. 2001. Catastrophic shifts in ecosystems. *Nature* 413:591–596.

Schimel, D. S. 1995. Terrestrial ecosystems and the carbon cycle. *Global Change Biology* 1:77–91.

Schimel, D. S., G. P. Asner, and P. Moorcroft. 2013. Observing changing ecological diversity in the Anthropocene. *Frontiers in Ecology and the Environment*, http://dx.doi.org/10.1890/120111.

Schimel, D. S., B. H. Braswell, and W. J. Parton. 1997. Equilibration of the terrestrial water, nitrogen, and carbon cycles. *Proceedings of the National Academy of Sciences of the United States of America*. 94:8280–8283.

Schimel, D. S., T. G. F. Kittel, A. K. Knapp, T. R. Seastedt, W. J. Parton, and V. B. Brown. 1991. Physiological interactions along resource gradients in a tallgrass prairie. *Ecology* 72:672–684.

Schimel, D., J. Melillo, H. Tian, A. D. McGuire, D. Kicklighter, T. Kittel, N. Rosenbloom, et al. Contribution of increasing CO2 and climate to carbon storage by ecosystems in the United States. 2000. *Science* 287:2004–2006.

Sellers, P. J., J. A. Berry, G. J. Collatz, C. B. Field, and F. G. Hall. 1992. Canopy reflectance, photosynthesis and transpiration, III. A reanalysis using enzyme kinetics-electron transport models of leaf physiology. *Remote Sensing of Environment* 42:187–216.

Shukla, J., C. A. Nobre, and P. J. Sellers, 1990. Amazonia deforestation and climate change. *Science* 247:1322–1325.

Sitch S., B. Smith, I. C. Prentice, A. Arneth, A. Bondeau, W. Cramer, J. Kaplan, et al. 2003. Evaluation of ecosystem dynamics, plant geography and terrestrial carbon cycling in the LPJ dynamic global vegetation model. *Global Change Biology* 9:161–185.

Steinberg, D. K., C. A. Carlson, N. R. Bates, R. J. Johnson, A. F. Michaels, and A. F. Knap. 2001. Overview of the U.S. JGOFS

Bermuda Atlantic Time-series Study (BATS): A decade-scale look at ocean biology and biogeochemistry. *Deep-Sea Research II* 48:1405–1447.

Stenseth, Nils Christian, Atle Mysterud, Geir Ottersen, James W. Hurrell, Kung-Sik Chan, and Mauricio Lima. 2002. Ecological effects of climate fluctuations. *Science* 269:1523–1528.

Sterner, R. W., and J. J. Elser. 2002. *Ecological stoichiometry: The biology of elements from molecules to the biosphere.* Princeton, NJ: Princeton University Press.

Steudler, P. A., R. D. Bowden, J. M. Melillo, and J. D. Aber. 1989. Influence of nitrogen fertilization on methane uptake in temperate forest soils. *Nature* 341:314–316.

Stocker, B. D., K. Strassmann, and F. Joos. 2011. Sensitivity of Holocene atmospheric CO_2 and the modern carbon budget to early human land use: Analyses with a process-based model. *Biogeosciences* 8:69–88.

Takahashi, T., J. Goddard, S. C. Sutherland, G. Mathiew, and D. W. Chipman. 1988. Assessment of carbon dioxide sink/source in the North Pacific Ocean: Seasonal and geographic variability, 1986–1987. Progress Report for Contract 19X-89675C, Oak Ridge National Laboratory, Oak Ridge, TN.

Tans, P. P., I. Y. Fung, and T. Takahashi. 1990. Observational constraints on the global atmospheric CO_2 budget. *Science* 247:1431–1439, doi:10.1126/science.247.4949.1431.

Tingley, M. W., W. B. Monahan, S. R. Beissinger, and C. Moritz. 2009. Colloquium papers: Birds track their Grinnellian niche through a century of climate change. *PNAS*: 106 Supplement 2: 19637–19643.

Vallis, G. K. 2011. *Climate and the oceans.* Princeton, NJ: Princeton University Press.

VEMAP. 1995. Vegetation/Ecosystem Modeling and Analysis Project (VEMAP): Comparing biogeography and biogeochemistry models in a continental-scale study of terrestrial ecosystem responses to climate change and CO2 doubling. *Global Biogeochemical Cycles* 9:407–437.

Vitousek, Peter M. 2004. *Nutrient cycling and limitation: Hawai'i as a model system*. Princeton, NJ: Princeton University Press.

Wang, Guiling, and David Schimel. 2003. Climate change, climate modes, and climate impacts. *Annual Review of Environment and Resources* 28:1–28.

Wilby, R. L., and T. M. L. Wigley. 1997. Downscaling general circulation model output: A review of methods and limitations. *Progress in Physical Geography* 21:530–548.

Wiley, E. O., Kristina M. McNyset, A. Townsend Peterson, C. Richard Robins, and Aimee M. Stewart. 2003. Niche modeling and geographic range predictions in the marine environment using a machine-learning algorithm. *Oceanography* 16:120–127.

Wolfram, Stephen. 2002. A new kind of science. http://www.wolframscience.com/nksonline/toc.html.

Woodwell, G. M., F. T. Mackenzie, R. A. Houghton, M. J. Apps, E. Gorham, and E. A. Davidson. 1995. Will the warming feed the warming? In *Biotic feedbacks in the global climatic system*, ed. G. M. Woodwell and F. T. Mackenzie, 393–411. New York: Oxford University Press.

Ziegler, C. L., T. J. Lee, and R. A. Pielke. 1997: Convective initiation at the dryline: A modeling study. *Monthly Weather Review* 125:1001–1026.

Index